Microelectromechanical Systems

Microelectromechanical Systems

Edited by **Eve Versuh**

New York

Published by NY Research Press,
23 West, 55th Street, Suite 816,
New York, NY 10019, USA
www.nyresearchpress.com

Microelectromechanical Systems
Edited by Eve Versuh

International Standard Book Number: 978-1-63238-323-5 (Hardback)

Printed in the United States of America.

Contents

Permissions

List of Contributors

Preface

This book is prepared by the specialists in the field of microelectromechanical systems (MEMS). It consists of two major sections dedicated to BioMEMS Devices, and MEMS characterization and micromachining. The developments of MEMS and devices have been influential in the demonstration of new devices and techniques, and even in the influx of new fields of research and development such as BioMEMS, actuators, microfluidic devices, RF and optical MEMS. The experience signifies a need for MEMS book encompassing these materials along with the most important process steps in bulk micro-machining and modeling. This book commences with the evolving field of bioMEMS, discussing MEMS coil for retinal prostheses, DNA extraction by micro/bio-fluidics devices and acoustic biosensors.

The researches compiled throughout the book are authentic and of high quality, combining several disciplines and from very diverse regions from around the world. Drawing on the contributions of many researchers from diverse countries, the book's objective is to provide the readers with the latest achievements in the area of research. This book will surely be a source of knowledge to all interested and researching the field.

In the end, I would like to express my deep sense of gratitude to all the authors for meeting the set deadlines in completing and submitting their research chapters. I would also like to thank the publisher for the support offered to us throughout the course of the book. Finally, I extend my sincere thanks to my family for being a constant source of inspiration and encouragement.

Editor

Part 1

BioMEMS Devices

MEMS-Based Microdevice for Cell Lysis and DNA Extraction

Xing Chen, Dafu Cui, Haoyuan Cai,
Hui Li, Jianhai Sun and Lulu Zhang
State Key Lab. of Transducer Tech., Inst. of Electronics, Chinese Academy of Sciences
China

1. Introduction

With the development of microelectromechanical system (MEMS) technology, micrototal analytical systems (μTAS) which has the potential for integrating sample pretreatment, target amplification, and detection, has been in progress. Micromachined analytical systems have several advantages over their large-scale counterparts, including low cost, disposability, low reagent and sample consumption, portability, and lower consumption. Many such devices have been demonstrated in the literature, including PCR microchips (Northrup et al., 1993; Copp et al., 1998; Panaro et al., 2005), DNA microchips (Fan et al., 1999), DNA biosensors (Kwakye et al., 2006), capillary electrophoresis (CE) microchips (Harrison et al., 1993; Backhouse et al., 2003; Liu et al., 2006), protein microchips (Yang et al., 2001; Wilson & Nie, 2006), etc. Most of these analytical processes need an effective yet simple method of obtaining high-quality DNA. Hence miniature devices for rapid sample pretreatment of DNA, including cell lysis and genomic DNA purification, are crucial for genetic application.

Traditional phenol extraction is a complex and time-consuming method for extracting DNA, and even some commercial purification kits require several centrifugal operations. The implementation of DNA purification on a microdevice is initially demonstrated based on the principle of solid phase extraction (SPE). The SPE on-microdevice can minimize sample loss and contamination problems as well as reduce analysis time, and besides, this SPE method can avoid problems of physical and biochemical degradation of DNA. For example, Tian et al. (Tian et al., 2000) established an SPE DNA purification microdevice in a capillary packing with silica resin matrix which could extract enough DNA for PCR reaction. Wolfe et al. (Wolfe et al., 2002) and Breadmore et al. (Breadmore et al., 2003) immobilized bare-silica beads matrix in microchannels by sol–gel technology for DNA purification. But a high packing density for larger surface area in the microfluidic device results in problems of backpressure and clogging of crude samples, and what is more, it is difficult to control the small particles in microdevices. A micropillar array fabricated by MEMS technology in a microchamber or channel increases the surface area available for DNA adsorption (Christel et al., 1999; Cady et al., 2003). However, the increasing surface area is limited and the problems of clogging could not be completely solved. Hence, a novel solid-phase matrix which should be easily integrated in microdevices is under demand.

It is well known that porous silicon with a relatively large specific surface area-to-volume (hundreds of square meters per cubic centimeter) can significantly increase the available interface area. Porous silicon can be easily obtained by electrochemical etching technology, and the pore geometry, the surface morphology, and the porosity of porous silicon can be very precisely controlled by the electrochemical etching conditions (Bisi et al., 2000; Schmuki et al., 2003). Porous silicon with several useful characteristics has already found applications in diverse fields such as solar cells (Bilyalov et al., 2003), RF (Park et al., 2001), (bio)chemical sensors (Massera et al., 2004; Björkqvist et al., 2004), etc.. Furthermore porous silicon with huge surface area as a kind of biocompatible solid support has been used to absorb enzyme, protein and other biologic molecules in fields of the enzyme micro reactors (Melander et al., 2005; Bengtsson et al., 2002), chromatography (Clicq et al., 2004), and antibody micro arrays (Steinhauer et al., 2005). Due to the fact that the technology for porous silicon fabrication is compatible with standard microelectronic and MEMS techniques, we have made use of porous silicon as the solid-phase matrix to extract DNA (Chen et al., 2006, 2007).

DNA purification from initial samples requires disrupting cells to liberate the nucleic acids before SPE process. Typical laboratory protocols for lysis steps include the use of enzymes (lysozyme), chemical lytic agents (detergents), and mechanical forces (sonication, bead milling). However, many such lysis techniques are not amenable to be implemented in a microfluidic platform. Miniaturization methods for lyzing cells are required so that cell lysis can be integrated into the μTAS. Therefore, a desirable lysis method should rapidly destroy cells while at the same time, it could not destroy nucleic acids and inhibit the following PCR reaction either, which would be amenable to be integrated with SPE on microdevices. Carlo et al. (Carlo et al., 2003) reported a mechanical cell lysis microdevice with nanostructural barbs which was used to disrupt sheep blood cells. However, the fabrication process of nanostructures is complex. Other miniaturization cell lysis methods include thermal (Lee & Tai, 1999), electrical (Gao et al., 2004; Taylor et al., 2001), ultrasonication (Belgrader et al., 2000; LaMontagne1 et al., 2002), and chemical treatments (Li & Jed Harrison, 1997). These approaches have been shown to be moderately successful. However, they all depend on the use of an external power supply and the devices may be quite complicated and costly to fabricate. Hence, chemical disruption methods are chosen due to the fact that they can be compatible with SPE on microdevices and do not require complex process of fabrication either. For example, Schilling et al. (Schilling et al., 2002) reported a relatively simple T-type microfluidic device that allowed the continuous lysis of bacterial cells using β-galactosidase. Sethu et al. (Sethu et al., 2004) established a continuous-flow microfluidic device for rapid erythrocyte lysis.

In this chapter, we have described a novel integrated MEMS-based microdevice capable of performing online cell lysis and genomic DNA purification during continuous flowing process. The method of chemical disruption was chosen for cell lysis, while the method of solid phase extraction was chosen for DNA purification. For cell lysis microdevice based on chemical disruption method, mixing is the key step due to the fact that each cell should be fully exposed to lysis buffer before the total lysis of cells, which is limited by the mixing speed. The mixing procedure was numerically simulated. Based on the results of the simulation, the mixing model for cell lysis was optimized to construct the microfluidic devices fabricated by MEMS technology. For DNA purification microdevice based on the SPE principle, solid-phase matrix is the key factor. The porous silicon was used as the solid-phase matrix to extract DNA. The porous silicon dioxide matrix was fabricated by electrochemical etching technology, and then were characterized by using SEM (scanning electron microscopy) and BET (Brunauer, Emmet, and Teller) nitrogen adsorption technique. The porous layer directly generated on the internal walls of the channels can greatly enhance the "active" surface area compared to the non-anodized one, and also avoid the problems of packing solid phase matrix.

In this integrated microdevice, cells were rapidly lyzed at first, and then genomic DNA was released and absorbed on a solid-phase matrix. Secondly, washing buffer was pumped through this microdevice for removing the proteins and other impurities which were also absorbed during the previous process. Finally, elution buffer was used to elute DNA and the desorbed DNA was collected at the outlet port. During continuous flow process, on-line rapid cell lysis and PCR amplifiable genomic DNA purification on the single MEMS-based microdevice have been implemented, which confirmed that the proposed microfluidic device was capable of directly providing genic analyte for the following molecular biology research or medical assay.

2. Theory

In microdevice channels, the flow is laminar in nature. Reynolds number, which is the ratio of the inertial forces to the viscous forces, can be used to characterize the flow. It is given by

$$Re = \rho \, v \, D_h \, / \, \mu \tag{1}$$

Where ρ is the fluid density, v is the flow velocity, and μ is the fluid viscosity. D_h represents the hydraylic diameter and, for rectangular channels, is given by

$$D_h = 2 \, a \, b / (a+b) \tag{2}$$

Where a and b are the cross sectional dimensions of the channel. The Reynolds numbers involved in the field of microfluidic device are usually below 100 (Gravesen et al., 1993), which prevent turbulent flow.

According the formula (1), the fluid density (ρ) and the fluid viscosity (μ) are constants. Re is depended on the flow velocity (v) and the hydraylic diameter (D_h). For our microdevice with a 200μm wide and 100μm deep rectangle channel, the hydraylic diameter (D_h) calculated by the formula (2) is 133μm. Note that the blood density (ρ) is 1.05g/cm³, and the viscosity of blood (μ) is 3.115mPa.s. According to the formula (1), the Re in our experimental was 0.035~8.4, when the flow velocity (v) is about 0.0008~0.024m/s. Therefore a typical laminar flow is found in our experimental. Lack of turbulent flow limits the effective mixing of the fluids under investigation. On the other hand, for the cell lysis on the microdevice based on the chemical method, it requires that each cell should be fully exposed to lysis reagents in microchannels, which is determined by mixing performance of the microdevice.

In microdevice channels where flow is laminar in nature, the mixing is dictated by diffusion. Average time for a small portion of a fluid to diffuse a distance L can be estimated by

$$T_D = L^2 \, / \, D \tag{3}$$

Where D is the diffusion coefficient of the liquid. The above equation can be used to predict the order of time scale of mass diffusion. As it suggests, one can dramatically reduce the mixing time by reducing the diffusion length required for mixing or increasing the contact area between two different liquids while keeping the volume constant.

3. Experimental

3.1 Reagents

Triton-X 100, HNO₃, KOH, ethanol, Tris, EDTA, and HCl were purchased from Beijing Chemical Reagents Company (Beijing China). SYBR Green I dye was purchased from Molecular Probes (Leiden, The Netherlands). Guanidine thiocyanate, λ-DNA, nucleic acid extraction kit, and PCR kit were purchased from Tianwei (Beijing China). Primers for the 203-bp, -gapd gene and primers for the 250-bp, -action gene were purchased from Sangon (Shanghai China).

Diluted buffer (0.9% NaCl in PBS) was used to dilute whole blood. TE buffer (10mM Tris, 1 mM EDTA, titrated to pH 8), load buffer (4M GuSCN in TE buffer with 1% Triton-X 100, titrated to pH 6.7), wash buffer (70% ethanol with 10mM NaCl) and eluted buffer (namely TE buffer) were used for the DNA purification procedure. All solutions were prepared in distilled water.

3.2 MEMS-based microdevice design and fabrication

T-type mixing model and sandwich type mixing model were proposed. In the sandwich type model, line-type model and coil-type model of the mixing microchannels were designed. All the models (see Fig.2) were numerically simulated. The best model with the best mixing performance was used to construct the silicon substrate. The mixing condition of the best model was optimized by numerical simulation, and then it was verified by cell lysis experiments.

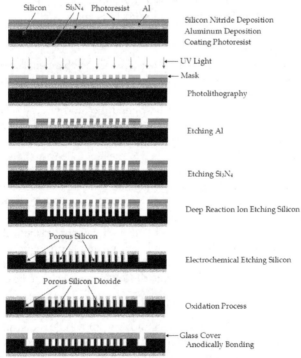

Fig. 1. Sequence for fabrication of the microdevice.

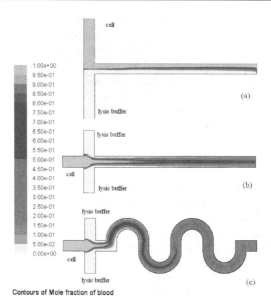

Contours of Mole fraction of blood

Fig. 2. Numerical simulation results for the species concentration distributions at T-type mixing model (a), sandwich type mixing model with lined channel (b) and sandwich type mixing model with coiled channel (c) when v_{cell} = 0.005m/s, v_{buffer} = 0.0025m/s. The color for "1.00e+00" and "0.00 e+00" in this figure refers to the concentration intensity of the pure cell sample and the pure lysis buffer, respectively. The software of FLUENT 6.2 was used to simulate.

The microdevice consisted of a silicon substrate and a glass cover. The fabrication process was shown in Fig. 1. A double-side polished n-type 0.01-0.1Ω·cm silicon wafer of (100) crystal orientation was first deposited with 0.3μm thick of silicon nitride (Si_3N_4) as the mask for fabricating porous silicon by a chemical vapour deposition (CVD) method, and then with 0.3μm aluminum as the mask for deep reaction ion etching (DRIE) by a electron beam vapour deposition technology. The wafer was spin coated with a positive photoresist (AZ1500) and patterned. After the exposed photoresist was developed, the exposed aluminum was removed by phosphoric acid and then the exposed silicon nitride was removed by plasma etching. The wafer was etched in a deep reaction ion etcher (Adixen, AMS100) to produce rectangle channels of 100μm in depth. After the process of removing the remained aluminum, a porous silicon layer on the internal walls of channels was anodized in 10% HF electrolyte at 20mA/cm² for 15min. After that, the silicon wafers were thoroughly rinsed in distilled water and oxidized at 1050°C for 1h. After oxidation process, the silicon nitride formerly patterned as the mask was removed by plasma etching for bonding. Finally glass covers (Corning Pyrex#7740), in which three holes had been drilled corresponding to the silicon substrate, were anodically bonded to the silicon wafers to form the closed channel by a bonder (Suss, SB6), thus microfluidic devices are fabricated.

3.3 Online cell lysis and DNA extraction procedure

The microdevice was mounted onto the stage of a microscope with a CCD camera and a video monitoring system. Blood was introduced through the cell inlet by a peristaltic pump, while lysis buffer (namely, load buffer) was pumped through the buffer inlet by another peristaltic pump. The value of the flow velocity was changed from 0.1μL/min to 25μL/min. When the lysis buffer and blood were blended in the channel during the continuous flow, cells were gradually lyzed online, and then genomic DNA was released and absorbed on the porous silicon matrix. DNA extraction procedure itself consisted of load, wash, and elution steps. In the load step, the released DNA was absorbed onto the porous silicon matrix in the presence of low pH (pH 6.7) and high concentration binding salt (4M GuSCN). And then proteins and possible PCR inhibitors were removed by passing 70% ethanol twice through the microdevice. Finally, DNA was eluted in TE buffer. 5 μL TE buffer was introduced in the microdevice for 10 min at 55°C and then 20 μL TE buffer was continuously passed through the microdevice.

3.4 Fluorescence detection and PCR amplification
DNA collected in the elution step was quantified by using SYBR Green I dye in a fluorometer using calibration curve of DNA which were generated using lambda DNA. Then DNA purified from whole blood was amplified by polymerase chain reaction (PCR). The 203-bp -gapd gene of rat's blood was amplified using the following primers: Forward Primer: 5'-AGAAGTACCTGCAACAGG-   3',   Reverse   Primer: 5'-GACGGACACATTGGGGGT- 3'. PCR reactions consisted of 2.5 μL standard 10×PCR buffer, 100 μmol dATP, dGTP, dCTP, and dTTP, 2.5 units of Taq polymerase, 50nmol of each primer, 5μL of the initial collected fraction, in a total volume of 25μL. These reactions were cycled in under the following conditions: 95°C denaturation for 5min, 35 cycles of 94°C for 1min, 68°C for 1min, 72°C for 1min, followed by a 10min extension at 72°C. DNA amplification was confirmed by gel electrophoresis.

4. Results and discussion

4.1 Design and simulation of mixing process
In general, comprehensive computational fluid dynamics (CFD) simulations are conducted before physical models are built and tested since these simulations enable the system parameters to be varied over a wide range of values and permit the simultaneous and instantaneous data collection of various aspects of models. Numerical simulation plays a key role in optimizing designs of microdevices and enables a reliable interpretation of the experimental results. T-type mixing model, sandwich-type mixing model with lined channel and coiled channel were designed and numerical simulated using FLUENT 6.2 software and all the simulations were two-dimensional.
The numerical simulation results are shown in Fig. 2. The mixing performance in sandwich-type mixing model is better than that in T-type one because of the increase of the contact area, and the mixing performance in coiled channel is much better than that in lined channel. Those numerical simulation results mentioned above provide a very clear understanding of the physical phenomena taking place in the two-dimensional microfluidic channels. In order to get a more accurate evaluation of the degree of mixing in the microchannel, a mixing index (σ) is used as follows (Erickson & Li, 2002):

$$\sigma = \left(1 - \frac{\int_0^h |c - c_\infty| dy}{\int_0^h |c_0 - c_\infty| dy}\right) \times 100\% \tag{4}$$

where C is the species concentration profile across the width of the microchannel (h), $c\infty$ is the completely mixed state (= 0.5) and C_0 is the completely unmixed state (= 0 or 1). Note that the confluent streams are completely mixed if σ = 100%. In contrast, they are completely unmixed if σ = 0%.

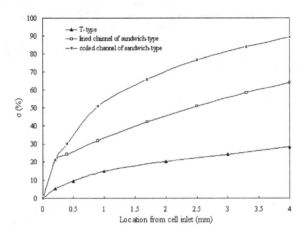

Fig. 3. Numerical evaluations of mixing efficiency index (σ) at different mixing models when v_{cell} = 0.005m/s, v_{buffer} = 0.0025m/s

Fig. 3 demonstrates the mixing efficiency of the proposed mixing models: T-type, sandwich-type with lined channel and sandwich-type with coiled channel. The use of coiled channel is effective in enhancing chaotic mixing, in which the mixing efficiency of 89.3% is obtained at the cross section located 4mm from inlet, by folding, stretching and reorienting fluid. Based on discussion above, the coiled channel of sandwich-type mixing model is chosen to construct the physical microdevice.

4.2 Optimization of cell lysis process

For cell lysis based on chemical disruption method, there are two steps which are mixing and chemical reaction respectively. In the microdevice, mixing is slow compared with the chemical reaction rate due to the fact that mixing in the microdevice is exclusively caused by diffusion because of laminar flow. The total lysis of cells requires that each cell should be fully exposed to lysis buffer, which is limited by the mixing speed. Thus the cell lysis efficiency is strongly depended on the mixing performance between the cell solution and the lysis buffer. From above results of simulation, a microdevice utilizing sandwich-type mixing model with coiled channel was designed and fabricated. As is shown in Fig. 4, the microdevice consists of a glass cover with two inlets and one outlet and a silicon substrate with an etched coiled channel of 200μm wide, 100μm deep and 20.16cm long.

The cell lysis performance is not only influenced by the design of the microchannels but also impacted by the velocity of the two solutions. With the increase of velocity, the time for cell lysis is reduced while the cell lysis efficiency decreases. In order to reduce the time for cell lysis and obtain high cell lysis efficiency, the velocity should be carefully examined. At first, the relationship between the mixing efficiency index and the inlet velocity was simulated using the coiled channel of sandwich-type mixing model.

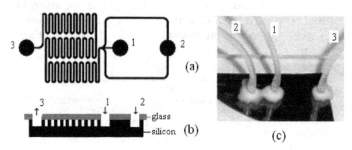

(a)

(b)

(c)

Fig. 4. Schematic top view (a), cross view (b) and photograph (c) of the microdevice.
(1) Cell inlet; (2) Buffer inlet; (3) Outlet.

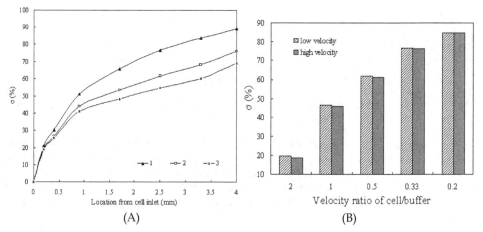

(A)   (B)

Fig. 5. Numerical evaluations of mixing efficiency index (σ) of coiled channel of sandwich-type mixing model. (Left) With different inlet velocities: "1" v_{cell} = 0.005m/s, v_{buffer} = 0.0025m/s; "2" v_{cell} = 0.02m/s, v_{buffer} = 0.01m/s; "3" v_{cell} = 0.2m/s, v_{buffer} = 0.1m/s. (Right) At the location near the inlet with different velocity ratio: "low velocity" v_{buffer} = 0.01m/s; "high velocity" v_{buffer} = 0.1m/s.

As is shown in Fig. 5A, the mixing efficiency is sharply decreased with the increase of the inlet velocity. This decrease of mixing efficiency can likely be attributed to the insufficient diffusion for the limited diffusion time. For the laminar flowing, the mixing time is proportional to the square of diffusion length so that decreasing diffusion length would reduce the mixing time sharply without deterioration of the mixing efficiency. The diffusion length can be changed by varying the cell and buffer velocity ratio. The lower cell/buffer velocity ratio leads to shorter diffusion length which can enhance mixing efficiency. Moreover at the location near the junction of three streams, the distribution of the laminar flow of the horizontal stream caused by the vertical streams enhances fluid mixing. And this enhancement is much stronger at the lower cell/ buffer velocity ratio. When the cell/buffer velocity ratio is 0.2, the mixing efficiency index is found to be more than 80% near the junction, which is insensitive to the value change of the flow velocity, shown in Fig. 5B.

In order to test the above results of simulation, the experiments of cell lysis were conducted by using the microfluidic device (Fig.4c). The cell sample was whole blood, while the lysis buffer was load buffer, 4M GuSCN in TE buffer with 1% Triton-X 100, titrated to pH 6.7. Triton X-100 is generic surfactants, with amphoteric chemical properties, which can react with water and can also react with lipid. High concentration guanidine salt is typical bonding salt for solid phase extraction nucleic acid and is also a generic strong denaturant. Moreover guanidine salt can quickly dissolve protein and can destroy the structure of cells. So the mixture of the traditional surfactant and the typical bonding salt was used as the lysis buffer for online cell lysis and DNA extraction.

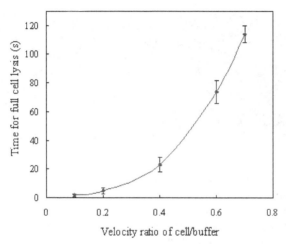

Fig. 6. Experimental evaluations of time for full cell lysis at variation of flow velocity ratio of cell/buffer when vbuffer = 1μL/min.

At first, the effect of flow velocity on the time taken for cell lysis was analyzed when the velocity ratio of cell/buffer was 1:1. For the flow velocity of 0.1, 0.2, 0.4 and 0.5μL/min, the time taken for complete lysis was tested to be 188s, 188s, 175s and 194s, at the location which was 3.92cm, 7.84cm, 14.56cm and 20.16cm from the inlet along the channel respectively. Note that the flow velocity of 1μL/min is equivalent to 0.833mm/s in the microchannel of 200μm wide and 100μm deep. The time taken for completely lysis at different flow rate with the same cell/buffer ratio was approximately the same; however cells could not be fully lyzed in the microdevice at more than 0.5μL/min. Then, the velocity ratio of cell/buffer was varied. As is shown in Fig. 6, the time required for full lysis is sharply reduced from approximate two minutes to several seconds with the decrease of the cell/buffer velocity ratio from 0.7 to 0.1. The photograph pictures of Fig. 7 show that the width of the cell stream is reduced with the decrease of the cell/buffer velocity ratio and is not affected by specific rate value. These experimental results are found to be consistent with the numerical results. Thus rapid cell lysis can be implemented under the condition of lower cell/buffer velocity ratio and higher specific rate value. In the end the cell velocity of 5μL/min and the lysis buffer velocity of 15~25μL/min were used for rapidly cell lysis.

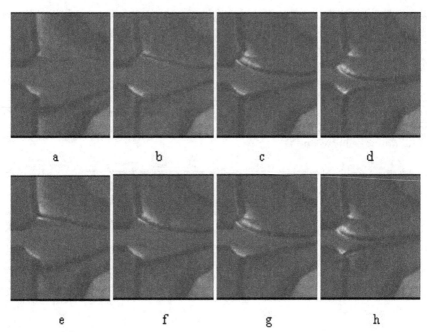

Fig. 7. Experimental results for the cell and buffer concentration distributions at location near the inlet with different velocity: "a" $v_{cell} : v_{buffer}$ = 2:1, v_{buffer} = 1μL/min; "b" $v_{cell} : v_{buffer}$ = 1:1, v_{buffer} = 1μL/min; "c" $v_{cell} : v_{buffer}$ = 1:3, v_{buffer} = 1μL/min; "d" $v_{cell} : v_{buffer}$ = 1:5, v_{buffer} = 1μL/min; "e" $v_{cell} : v_{buffer}$ = 2:1, v_{buffer} = 10μL/min; "f" $v_{cell} : v_{buffer}$ = 1:1, v_{buffer} = 10μL/min; "g" $v_{cell} : v_{buffer}$ = 1:3, v_{buffer} = 10μL/min; "h" $v_{cell} : v_{buffer}$ = 1:5, v_{buffer} = 10μL/min.

4.3 DNA purification using porous silicon matrix

DNA purification on-microdevice is based on SPE method: firstly adsorption of DNA onto a solid-phase matrix surface and then washing the impurities such as protein and finally elution of the purified DNA. The mechanism of DNA adsorption on a solid-phase matrix surface elucidated by Melzak et al. (Melzak et al., 1996) was that the adsorption of highly charged duplex DNA to negatively charged silica was controlled by three competing effects: weak electrostatic repulsion forces, dehydration and hydrogen bond formation. The surface area of the matrix significantly affects the adsorption of DNA onto the solid-phase matrix surface.

Porous silicon gotten on the surface of the tortuous rectangle channel in the microdevice by electrochemical etching technology can enhance the available surface area (see Fig. 8). The details of the preparation and optimization process for porous silicon were reported in our previous research (Chen et al., 2006; Chen et al., 2007). By using BET nitrogen adsorption experiment, pore size of the porous silicon is in the range of 20nm to 30nm. And the surface area of porous silicon is approximate 400m²/g. Note that the volume of microfluidic device with a tortuous channel of 25cm long, 100μm deep and 200μm wide is 0.005cm³. Thus the surface area to volume of microdevice with porous silicon is approximate 300m²/cm³ which is thousands of times higher than that of one without porous silicon since it is known that the surface area to volume of the microdevice without porous silicon is approximate 200cm²/cm³.

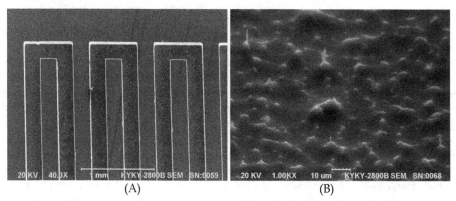

Fig. 8. SEM of porous silicon on the surface of the channel for DNA purification. (A) ×40; (B) ×1000.

The microdevices with and without porous silicon were used to recover DNA from 200ng prepurified genomic DNA under the same experimental conditions, respectively. The performance of the microdevice with porous silicon is quite well, with an average of 83% (11.6%RSD) evaluated from five extractions, which is much higher than that of one without porous silicon with an average of 39.2% (8.7%RSD). It is important to note that 200ng DNA overloads the capacity of the microdevice without porous silicon due to the limitation of surface area. Therefore the binding capacity of the microdevice without porous silicon is approximately 75ng/cm² which agrees with the results of Cady et al. (Cady et al., 2003) who found that the binding capacity of nonporous micropillars was approximate 82ng/cm². The previous researches proved that the performance of DNA extraction microdevice was determined by the surface area of the matrix and the extracted DNA was found to increase linearly with the surface area (Fan et al., 1999; Kwakye et al., 2006). The binding capacity of porous silicon matrix would increase hundreds or thousands of times, while the extraction efficiency didn't improve so much. The reason could probably be that most of the internal pores and smaller pores might not be used to adsorb DNA, and the DNA adsorbed in these pores might not be easily eluted.

Comparison with the microdevice without the porous silicon layer, the microdevice with the porous layer achieved higher extracting efficiency of DNA without problems of clogging and high backpressure. Moreover this SPE microdevice with porous silicon can be easily integrated with other microdevices, since the porous silicon can be directly generated on the internal walls of channel in the microdevice.

4.4 Cell lysis and purification of DNA on microdevice

A microfluidic device capable of real time destroying cells and extracting DNA from the cell lysates during continuous flow was designed and fabricated. As is shown in Fig. 4, the microfluidic device with the size of 2cm×1.2cm consists of a silicon substrate within an etched coiled channel, and a glass cover with two inlets and one outlet according to the silicon substrate. The two inlets: cell inlet and buffer inlet are designed to introduce blood sample and lysis buffer simultaneously, leading to rapid lysis of blood cells near the location of cell inlet using the sandwich-type mixing model. Porous silicon layer fabricated on the surface of the internal wall of the coiled channel is designed to absorb DNA, which can strongly improve the extraction efficiency with the huge surface area. Thus this microfluidic device makes it possible to lyze cells and to purify DNA orderly.

Microfluidic devices with porous silicon matrix were used for lyzing blood cells and purifying genomic DNA, including three steps: load step, washing step, and elution step. In load step, blood cells were damaged, and the genomic DNA was released and adsorbed on the porous silicon matrix. In washing step, impurity such as protein was removed while the genomic DNA was still held on the porous matrix. In elution step, the purified DNA was desorbed from the porous matrix into the elution buffer. The process of online cell lysis and DNA extraction was in details as follows. Firstly, 10μL whole blood was introduced into the channel from the cell inlet, while 50μL lysis buffer was introduced from the buffer inlet. The lysis buffer was load buffer, namely 4M GuSCN in TE buffer with 1% Triton-X 100, titrated to pH 6.7, which included lysis reagent and bonding salt for DNA adsorption. GuSCN and Triton-X 100 were both the chemical reagents for cell lysis, while GuSCN is the bonding salt for DNA extraction. The flow velocity of whole blood sample was 5μL/min, whereas the flow velocity of lysis buffer was 25μL/min. The whole blood sample and the lysis buffer were mixed effectively at the location near the inlet, and blood cells were rapidly lyzed. After cell lysis, genomic DNA in blood cells was released and adsorbed on the porous matrix at the presence of low pH (pH 6.7) and high concentration binding salt (4M GuSCN). Secondly, 50μL washing buffer was pumped in the microchannel at 25μL/min from the buffer inlet, while the cell inlet was closed during washing step. Finally, 10μL of TE buffer was introduced in the microdevices and incubated in the microdevices for 5min at 55°C and then another 50μL of TE buffer was passed through continuously. As a whole, online cell lysis and genomic DNA purification could be implemented in less than 20min by using this integrated microdevice, while its large-scale counterparts commonly require more than several hours to finish the job.

In order to value the performance of the integrated microdevice, the elution buffer flowing out from the outlet port, in which the purified DNA was desorbed and eluted, was collected and subjected to fluorescence detection using SYBR Green I. From fluorescence detection, average 39.7ng genomic DNA was extracted from 1μL whole blood with 10.5%RSD by using three microdevices, respectively, at the same conditions. The extracted efficiency of DNA by the integrated microdevices is higher than commercial kit with silica resin which can only extract about 20~30ng DNA per microlitre blood. Moreover, it is only 10μl blood sample needed for the integrated microdevice to extract genomic DNA, while traditional phenol extraction or commercial kit requires several milliliters blood because of centrifugal operation.

The extraction of genomic DNA from a crude biological sample must be PCR-amplifiable. The lyzed cells are a complex mixture of proteins, peptides, lipids, carbohydrates, and other low molecular weight compounds that are known to inhibit DNA amplification by PCR. The genomic DNA extracted from whole blood in two separate experiments, respectively, was submitted for PCR to ensure that no inhibitory compounds were present. Shown in Fig. 9, a 203-bp fragment of -gapd gene is successfully amplified, which is identified by gel electrophoresis separation. This illustrates that the eluted DNA using the microdevice has been purified effectively for subsequent enzymatic reactions. Also shown from Fig. 9, the repeatability of this system is good enough for the downstream analytical steps.

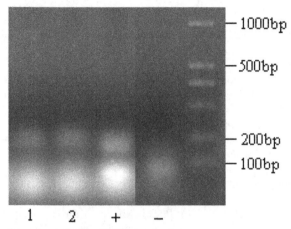

Fig. 9. Gel electrophoresis analysis of PCR products.

5. Conclusion

This chapter has demonstrated an integrated microdevice capable of performing online cell lysis and DNA extraction. Firstly micro total analytical systems (μTAS) base on MEMS technology was briefly introduced in Section 1. Then microdevices both for DNA extraction and cells lyzing were reviewed. The theory of flowing in microdevice was introduced in Section 2. In microdevice channels, the flow is laminar in nature. Then a novel MEMS-based microdevice capable of performing online cell lysis and DNA extraction was reported in Section 3 and 4. The fabrication procedure of the MEMS-based microdevice was presented in detail, while the experimental of cell lysis and DNA extraction were described respectively. Then the properties of this microdevice were studied by numerical simulation and experimental. According to the results of the simulation, the mixing performance in sandwich-type coiled channels was better than that in the T-type ones. According to the results of the experimental, a high performance of cell lysis was obtained in the sandwich-type microdevice at the optimized conditions of the cell/buffer velocity ratio <1/5. 83% DNA was recovered by the porous matrix, strongly contrary to the 39.2% DNA extracted by the non-porous one. And 39.7ng PCR-amplifiable genomic DNA was obtained from whole blood samples by using an integrated microdevice with sandwich-type coiled channels and porous silicon matrix. In general, the developed integrated microdevice providing a powerful tool for biological sample pre-treatment is shown to exhibit numerous advantages over its large-scale counterparts, including rapidness, much lower blood and reagent consumption, disposability, and portability and so on, which has the potential to integrate into μTAS for point-of -care medical diagnose.

6. Acknowledgement

The authors greatly acknowledge the financial support from the National Science Foundation of China under Grant number 60701019, 60427001 and 60501020. The authors are grateful to Mr. Feng Shen of Institute of Mechanics, Chinese Academy of Sciences for his assistance.

7. References

Backhouse, C.J.; Gajdal, A.; Pilarski, L.M. & Crabtree, H.J. (2003). Improved Resolution with Microchip-based Enhanced Field Inversion Electrophoresis, *Electrophoresis*, Vol.24, pp.1777-1786

Belgrader, P.; Okuzumi, M.; Pourahmadi, F.; Borkholder, D.A. & Northrup, M.A. (2000). A Microfluidic Cartridge to Prepare Spores for PCR Analysis, *Biosens. Bioelectron.*, Vol.14, pp.849–852.

Bengtsson, M.; Ekström, S.; Marko-Varga, G. & Laurell, T.; (2002). Improved Performance in Silicon Enzyme Microreactors Obtained by Homogeneous Porous Silicon Carrier Matrix, *Talanta*, Vol.56, pp.341-53

Bilyalov, R.; Stalmans, L.; Beaucarne, G.; Loo, R.; Caymax, M.; Poortmans, J.; & Nijs, J. (2001). Porous Silicon as an Intermediate Layer for Thin-Film Solar Cell, *Solar Energy Materials and Solar Cells*, Vol.65, No.1-4, pp.477-85

Bisi, O.; Ossicini, S. & Pavesi, L. (2000). Porous Silicon: a Quantum Sponge Structure for Silicon based Optoelectronics, *Surface Science Reports*, Vol.38, pp.1-126

Björkqvist, M.; Salonen, J.; Paski, J.; & Laine, E. (2004). Characterization of Thermally Carbonized Porous Silicon Humidity Sensor, *Sensors and Actuators A*, Vol.112, pp.244-7

Breadmore, M.C.; Wolfe, K.A.; Arcibal, I.G.; Leung, W.K.; Dickson, D.; Giordano, B.C.; Power, M.E.; Ferrance, J.P.; Feldman, S.H.; Norris, P.M. & Landers, J.P. (2003). Microchip-based Purification of DNA from Bilological Samples, *Anal. Chem.*, Vol.75, pp.1880-1886.

Cady, N.C.; Stelick, S. & Batt, C.A. (2003). Nucleic Acid Purification Using Microfabricated Silicon Structures, *Biosens. Bioelectron.*, Vol.19, pp.59-66.

Carlo, D.D.; Jeong, K.H. & Lee, L.P. (2003). Reagentless Mechanical Cell Lysis by Nanoscale Barbs in Microchannels for Sample Preparation, *Lab chip*, Vol.3, pp.287-291.

Chen, X.; Cui, D. F. Liu, C. C. & Li, H. (2007). Microfabrication and Characterization of Porous Channels for DNA Purification, *J. Micromech. Microeng.*, Vol.17, pp.68-75

Chen, X.; Cui, D. F.; Liu, C. C. & Cai, H. Y. (2006). Fabrication of Solid Phase Extraction DNA Chips based on Bio-Micro-electron-Mechanical System Technology, *Chin. J. Anal. Chem.*, Vol.34, pp.433−436

Christel, L.A.; Petersen, K.; McMillan, W. & Northrup, M.A. (1999). Rapid Automated Nucleic Acid Probe Assays Using Silicon Microstructures for Nucleic Acid Concentration, *Transactions of the ASME*, Vol.121, pp.22-27.

Clicq, D.; Tjerkstra, R. W.; Gardeniers, J. G. E.; van den Berg, A.; Baron, G. V. & Desmet, G.; (2004). Porous Silicon as a Stationary Phase for Shear-Driven Chromatography, *Journal of Chromatography A*, Vol.1032, pp.185-91

Copp, M.U.; Luechinger, M.B. & Manz, A. (1998). Chemical Amplification: Continuous-Flow PCR on a Chip, *Science*, Vol.280, pp.1046-1048

Erickson, D. & Li, D. (2002). Influence of Surface Heterogeneity on Electrokenetically Driven Microfluidic Mixing, *Langmuir*, Vol.18, pp.1883-1892.

Fan, Z.H.; Mangru, S;. Granzow, R.; Heaney, P.; Ho, W.; Dong, Q. & Kumar, R. (1999). Dynamic DNA Hybridization on a Chip Using Paramagnetic Beads, *Anal. Chem.*, Vol.71, pp.4851-4859

Gao, J.; Yin, X.F. & Fang, Z.L. (2004). Integration of Single Cell Injection, Cell Lysis, Separation and Detection of Intracellular Constituents on a Microfluidic Chip, *Lab chip*, Vol. 4, pp.47-52.

Gravesen, P.; Branebjerg, J. & Jensen, O.S. Microfluidics-a Review, (1993). *J. Micromech. Microeng.*, Vol.3, pp.168-182.

Harrison, D.J.; Fluri, K.; Seiler, K.; Fan, Z.; Effenhauser, C.S. & Manz, A. (1993). Micromaching a Miniaturized Capillary Electrophoresis-based Chemical Analysis System on a Chip, *Science*, Vol.261, pp.895-897

Kwakye, S.; Goral, V.N. & Baeumner, A.J. (2006). Electrochemical Microfluidic Biosensor for Nucleic Acid Detection with Integrated Minipotentiostat, *Biosens. Bioelectron.*, Vol.21, pp.2217–2223

LaMontagne1, M.G.; Jr., F.C.M.; Holden, P.A. & Reddy, C.A. (2002). Evaluation of Extraction and Purification Methods for Obtaining PCR-Amplifiable DNA from Compost for Microbial Community Analysis, *J. Microbiol. Meth.*, Vol.49, pp.255–264.

Lee, S.W. & Tai, Y.C. (1999). A Micro Cell Lysis Device, *Sens. Actuat. A*, Vol.73, pp.74–79.

Li, P.C.H.& Jed Harrison, D. (1997). Transport, Manipulation, and Reaction of Biological Cells On-Chip Using Electrokinetic Effects, *Anal. Chem.*, Vol.69, pp.1564-1568.

Liu, D.Y.; Shi, M.; Huang, H.Q.; Long, Z.C.; Zhou, X.M; Qin, J.H. & Lin, B.C. (2006). Isotachophoresis Preconcentration Integrated Microfluidic Chip for Highly Sensitive Genotyping of the Hepatitis B Virus, *J. Chromatogr. B*, Vol.844, pp.32-38

Massera, E.; Nasti, I.; Quercia, L.; Rea, I.; & Di Francia, G. (2004). Improvement of Stability and Recovery Time in Porous-Silicon-based NO_2, *sensor Sensors and Actuators B*, Vol.102, pp.195-7

Melander, C.; Bengtsson, M.; Schagerlöf, H.; Tjerneld, F.; Laurell, T. & Gorton, L. (2005). Investigation of Micro-Immobilised Enzyme Reactors Containing Endoglucanases for Efficient Hydrolysis of Cellodextrins and Cellulose Derivatives, *Analytica Chimica Acta*, Vol. 550, pp.182-90

Melzak, K.A.; Sherwood, C.S.; Turner, R.B. & Haynes, C.A. (1996). Driving Forces for DNA Adsorption to Silica in Perchlorate Solutions, *J. Colloid Interf. Sci.*, Vol.181, pp.635–644.

Northrup, M.A.; Ching, M.T.; White, R.M. & Watson, R.T. (1993). DNA Amplification with a Microfabricated Reaction Chamber,*Transducer*, pp.924-926

Panaro, N.J.; Lou, X.J.; Fortina, P.; Kricka, L.J. & Wilding, P. (2005). Micropillar Array Chip for Integrated White Blood Cell Isolation and PCR, *Biomolecular Engineering*, Vol.21, pp.157–162

Park, J. Y. & Lee, J. H. (2003). Characterization of 10 μm Thick Porous Silicon Dioxide Obtained by Complex Oxidation Process for RF Application, *Materials Chemistry and Physics*, Vol.82, pp.134-9

Schilling, E.A.; Kamholz, A.E. & Yager, P. (2002). Cell Lysis and Protein Extraction in a Microfluidic Device with Detection by a Fluorogenic Enzyme Assay, *Anal. Chem.*, Vol.74, pp.1798-1804.

Schmuki, P.; Schlierf, U.; Herrmann, T. & Champion, G. (2003) Pore Initiation and Growth on n-InP(100), *Electrochimica Acta*, Vol.48, pp.1301-8

Sethu, P.; Anahtar, M.; Moldawer, L. L.; Tompkins, R. G. & Toner, M. (2004). Continuous Flow Microfluidic Device for Rapid Erythrecyte Lysis, *Anal. Chem.*, Vol.76, pp.6247–6253.

Steinhauer, C.; Ressine, A.; Marko-Varga, G.; Laurell, T.; Borrebaeck, C. A. K. & Wingren, C. (2005). Biocompatibility of Surfaces for Antibody Microarrays: Design of Macroporous Silicon Substrates, *Analytical Biochemistry*, Vol.341, pp.204-13

Taylor, M.T.; Belgrader, P.; Furman, B.J.; Pourahmadi, F.; Kovacs, G.T.A. & Northrup, M.A. (2001). Lysing Bacterial Spores by Sonication through a Flexible Interface in a Microfluidic System, *Anal. Chem.*, Vol.73, pp.492-496.

Tian, H;. Hühmer, A.F.R. & Landers, J.P. (2000). Evaluation of Silicon Resins for Direct and Efficient Extraction of DNA from Complex Biological Matrices in a Miniaturized Format, *Anal. Biochem.*, Vol.283, pp.175-191.

Waters, L.C.; Jacobson, S.C.; Kroutchinina, N.; Khandurina, J.; Foote, R. S. & Ramsey, J.M. (1998). Microchip Device for Cell Lysis, Multiplex PCR Amplification, and Electrophoretic Sizing, *Anal. Chem.*, Vol.70, pp.158-162.

Wilson, M.S. & Nie, W. (2006). Multiplex Measurement of Seven Tumor Markers Using an Electrochemical Protein Chip, *Anal. Chem.*, Vol.78, pp.6476-6483

Wolfe, K.A.; Breadmore, M.C.; Ferrance, J.P.; Power, M.E.; Conroy, J.F.; Norris, P.M. & Landers, J.P. (2002). Toward a Microchip-based Solid-Phase Extraction Method for Isolation of Nucleic Acids, *Electrophoresis*, Vol.23, pp.727–733.

Yang, T.; Jung, S.; Mao, H. & Cremer, P.S. (2001). Fabrication of Phospholipid Bilayer-Coated Microchannels for on-Chip Immunoassays, *Anal. Chem.*, Vol.73, pp.165-169

Implantable Parylene MEMS RF Coil for Epiretinal Prostheses

Wen Li[1], Damien C. Rodger[2], James D. Weiland[2],
Mark S. Humayun[2], Wentai Liu[4] and Yu-Chong Tai[3]
[1]Michigan State University,
[2]University of Southern California,
[3]California Institute of Technology
[4]University of California, Santa Cruz
USA

1. Introduction

Age related macular degeneration (AMD) and retinitis pigmentosa (RP) are two of the most common outer retinal degenerative diseases that have resulted in vision impairment and blindness of millions of people. Specifically, AMD has become the third leading cause of blindness on global scale, and is the primary cause of visual deficiency in industrialized countries (World Health Organization [WHO], 2011). In the United States, more than 500,000 people are suffering from RP and around 20,000 of them are legally blind (Artificial Retina Project, 2007). Whereas many treatment methods, including gene replacement therapy (Bennett et al., 1996), pharmaceutical therapy, nutritional therapy (Norton et al., 1993), photoreceptor and stem cell transplantations (MacLaren et al., 2006 & Tropepe et al., 2000), and dietary, have been explored to slow down the development of AMD and RP diseases in their early stages, vision impairment and blindness due to outer retinal degeneration currently remain incurable.

In the 1990's, researchers discovered that although the retinal photoreceptors are defective in RP patients, their optic nerves, bipolar and ganglion cells to which the photoreceptors synapse still function at a large rate (Humayun et al., 1999). Further studies showed similar results in AMD patients (Kim et al., 2002). These findings have created profound impact on the ophthalmology field, by providing the possibility of using artificial retinal prostheses to partially restore the lost vision function in AMD and RP patients. Two main retinal implant approaches are currently in development according to the layer of retina receiving the implanted device: subretinal (Chow et al., 2006; Rizzo, 2011; Zrenner et al., 1999) and epiretinal prostheses (Humayun et al., 1994; Weiland & Humayun 2008; Wong et al., 2009). Particularly, epiretinal implantation has received widespread attention over the last few years, for not only successful clinical trials demonstrating its efficacy in patients, but also its many advantages compared to others (Horch K.W. & Dhillon, G.S., 2004). First, the device implantation and follow-up examination only require standard ophthalmologic technologies, which can effectively reduce the risk of trauma during the surgery and also allow for the implant to be replaced easily. In addition, most of the implanted electronics are kept in the vitreous cavity so that potential thermal damage to the surrounding retinal tissues can be mitigated by taking the advantage of the heat dissipating properties of the vitreous. Finally, developments of modern microelectromechanical systems (MEMS) technologies enable complete intraocular retinal implants and high-density array stimulation.

Various epiretinal prosthetic technique options have been studied by a number of groups worldwide, by replacing the defective photoreceptors with a multielectrode array implanted on the surface of the inner retina between the vitreous and internal limiting membrane (Javaheri et al. 2006; Rizzo et al., 2004; Stieglitz et al. 2004). Although they have different system configurations, these systems generally utilize a pair of coils to transfer data and power wirelessly between an extraocular data acquisition system and intraocular electronics. In such implementations, the intraocular receiver coil, which resides inside the eye, has many constrains compared to the extraocular transmitter coil. First, the coil has to be mechanically durable to withstand the surgical procedure. It also needs to be flexible and small enough to facilitate the device implantation through small incisions. Finally, it should be chemically inert and biocompatible to prevent harmful interaction with surrounding tissue/cells. Currently, most systems are still using thick and stiff hand-wound coils as the receiver coils, which can cause notable degradation in the implant region. In addition, interconnections between the hand-wound coils and other components are usually formed by soldering, which will require additional hermetic package to ensure device biocompability. Planar microcoils that contain electroplated gold wires on flexible polyimide substrates have also been developed (Mokwa et al. 2008). While device functionality has been confirmed in animal models, several critical issues such as long-term reliability and biocompatibility need to be further addressed for human implantation.

To overcome these challenges, we have proposed a polymer-based MEMS technique for making microcoils. Devices typically consist of multi-layer conductive wires encapsulated by polymeric materials (Li et al., 2005; Li et al., 2006; Chen et al., 2008). In this design, Parylene C is selected as a main structural and packaging material because of its many unique properties, such as flexibility (Young's modulus ~ 4 GPa), chemical inertness, United States Pharmacopoeia (USP) Class VI biocompatibility (Rodger et al., 2006), and lower water permeability compared with other commonly used polymers (e.g. PDMS and polyimide) (Licari & Hughes, 1990). Microfabrication technology has several advantages over the hand-winding including miniaturized structures, precise dimensional control, and feasibility for system integration. The proposed coils will finally be integrated with other system components, such as circuitry, discrete electrical components, a flexible cable, and a high-density multielectrode array, to achieve a complete all-intraocular epiretinal prosthetic system (Rodger et al., 2008), as described in Fig. 1.

This chapter will concentrate on the design, microfabrication, and testing results of two types of Parylene-based intraocular MEMS coils for applications in epiretinal implantation. Specifically, Section 2 will introduce theoretical approaches for coil modeling and design. Coil's electrical properties, including self-inductance, effective series resistance (ESR), and parasitic capacitance, will be discussed theoretically with respect to their geometries. Section 3 describes a Parylene-metal-Parylene thin film technology for making the proposed microcoils. Two various types of microfabricated coil prototypes, including regular dual-layered planar coils and novel fold-and-bond coils, are implemented and characterized in Section 4. Successful data transmissions through such devices have been demonstrated using inductive coupling tests. Experiment also confirms that the quality factor (Q factor) of microcoils can increase proportionally with the increase of the number of metal layers.

Preliminary results suggest that the fold-and-bond technology is a very promising approach for making high Q MEMS coils in a simple, low-cost, and microfabrication compatible manner. Future direction for coil optimization will aim to increase the number of metal layers in order to enhance the Q factor and the power transfer efficiency.

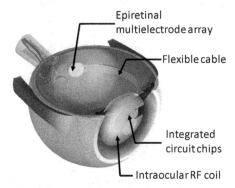

Fig. 1. System schematic of an all-intraocular retinal prosthetic system, which contains two MEMS radio-frequency (RF) coils for power and data transmission, circuitry integrated on a flexible Parylene cable for converting the signal to simulation pulses, and a high-density MEMS electrode array for simulating the neural cells.

2. Modeling of microcoils

Microfabricated coils usually suffer from low self-inductances and inevitable parasitic effects, namely parasitic resistances and capacitances, due to their small physical dimensions and technical constrains of surface micromachining. Particularly, for intraocular retinal implants in human subjects, planar coils with a maximal outer diameter of ~ 10 mm and a minimal inner diameter of ~ 3 mm are desired, which is limited by the space availability in the anterior chamber of human eyes. To better understand the electrical properties and parasitic effects in such small devices, we studied several analytical models and have now applied them to a simplified circular-shape planar coil as illustrated in Fig. 2. In this section, the coil self-inductance, ERS, and parasitic capacitance will be discussed separately with respect to the geometric parameters of devices.

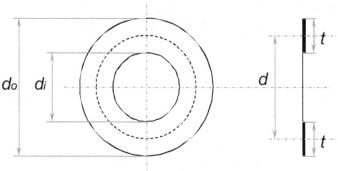

Fig. 2. Simplified model of a circular-shape planar MEMS coil.

2.1 Self-inductance

The self-inductance (L_s) of a multi-layer circular coil with rectangular cross-section can be calculated using the following equation (Dwight , 1945):

$$L_s = 2\pi d(nN)^2 \times 10^{-9} \left[\left(\ln\frac{4d}{t} \right) \left(1 + \frac{t^2}{24d^2} \cdots \right) - \frac{1}{2} + \frac{43t^2}{288d^2} \cdots \right] \quad \text{(in Henry)}, \tag{1}$$

where d (in cm) is the mean diameter of the coil, t (in cm) is the coil width, n is the number of turns on each layer, and N is the total number of layers. This expression is valid only when the coil is operated at a low-frequency, i.e., no skin effect. The skin effect can be evaluated using a frequency-dependent factor, which is known as skin depth δ and can be calculated as

$$\delta = \sqrt{\frac{2\rho}{\omega\mu}} \quad \text{(in meter)}, \tag{2}$$

where ρ is the electrical resistivity of metal (in Ω·m), ω is the angular frequency (in rad/s), and μ is the permeability of metal (in H/m). In our proposed system, the data signal is modulated on a ~ 22 MHz carrier, and the power transfer is taken place at a frequency within 1-2 MHz. Therefore, the estimated skin depths at these low frequencies are much bigger than the thickness of metal thin films produced from physical vapor deposition (PVD). In this case, the skin effect can be negligible with an assumption of uniform current distribution in conductive wires.

2.2 Effective series resistance

ERS (R_s) is commonly used to estimate coil losses, which plays an important role in designing a power efficient inductive link. The ESR can be divided into two parts: DC resistance and frequency dependent resistance. Assuming the width of metal traces is much larger than the separation distance between adjacent turns that can be ignored, the DC resistance of the proposed coil can be calculated with the Ohm's law as given in (3),

$$R_s = \rho\frac{n^2 N\pi d}{t \times h} \quad (\Omega), \tag{3}$$

where ρ is the metal resistivity (in Ω·m), and h is the metal thickness (in m). As a result of the skin-effect, the frequency dependent part can be neglected at the low operating frequencies as mentioned earlier. Therefore, the equivalent ESR can be simply written as a DC resistance.

2.3 Parasitic capacitance

Parasitic capacitance (C_s) places a limit to the self-resonant frequency of the coil, above which the coil will not behave as an inductor any more. In a first-order approximation, the parasitic capacitance of a planar MEMS coil usually consists of two main components: the capacitance between turns and the capacitance between layers. A distributed model has been developed to estimate the equivalent parasitic capacitance, as discussed elsewhere (Wu et al. 2003 & Zolfaghari et al. 2001). In this method, a planar coil can be decomposed into equal sections by assuming consistent thickness and width of metal traces everywhere. The voltage profile can then be obtained by averaging the beginning and ending potential across the coil structure. With known voltage variations between the correlated sections of adjacent turns and layers, the total capacitive energy stored in the coil structure can be calculated using the distributed capacitance of each segment. The equivalent capacitance can then be approximated from the distributed capacitances, using the ideal double plate capacitor formula.

$$C_{eq-turn} = \sum_{k=1}^{n-1} \frac{1}{4} C_{ii} l_k [d(k+1) - d(k-1)]^2, \tag{4}$$

$$C_{eq-layer} = \frac{1}{4} \sum_{k=1}^{N} (C_{m,m-1} + C_{m-2,m-3} + ...) \frac{A_k}{m^2} [4 - 2d(k-1) - 2d(k)]^2$$
$$+ \frac{1}{4} \sum_{k=1}^{N} (C_{m-1,m-2} + C_{m-3,m-4} + ...) \frac{A_k}{m^2} [2d(k-1) + 2d(k)]^2, \tag{5}$$

$$C_{eq-total} = C_{eq-turn} + C_{eq-layer}, \tag{6}$$

Equations (4), (5), and (6) show the analytical formulas for calculating parasitic capacitance, where C_{ii} (in F) denotes the capacitance per unit length between adjacent metal turns, $C_{m,m-1}$ (in F) is the capacitance per unit area between the m-th and (m-1)-th metal layer, A_k (in m²) is the trace occupied area of the k-th turn on each layer, and $d(k) = h_1 + h_2 + ... + h_k$, in which h_k is defined as the ratio of the wire length of the k-th turn (l_k) to the total wire length (l_{tot}). This simplified model neglects the second order parasitic capacitances between non-adjacent turns and layers, which are much less than the first order capacitances.

2.4 Quality factor

Q factor is an important metric for evaluating the efficiency of a coil, which is theoretically defined as the ratio of total stored energy to dissipated energy per cycle in a resonating system. With known L_s, R_s, and C_s, the Q factor of the coil can be derived from a 3-element circuit model (Fig. 3) (Wu, 2003). In order to obtain the coil Q factor mathematically, the total equivalent impedance (Z_s) is first studied, which can be written as the sum of a real resistance and an imaginary reactance (equation [7]).

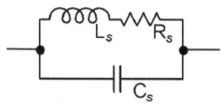

Fig. 3. Equivalent RLC circuit of a planar MEMS coil.

$$Z_s = \frac{R_s}{(1 - \omega^2 L_s C_s)^2 + (\omega C_s R_s)^2} + j \frac{\omega(L_s - R_s^2 C_s - \omega^2 L_s^2 C_s)}{(1 - \omega^2 L_s C_s)^2 + (\omega C_s R_s)^2}. \tag{7}$$

Then the self-resonant frequency ω_s can be expressed as:

$$\omega_s = \sqrt{\frac{1}{L_s C_s} - \frac{R_s^2}{L_s^2}} \approx \sqrt{\frac{1}{L_s C_s}}, \text{ when } R_s \ll \sqrt{\frac{L_s}{C_s}}. \tag{8}$$

For a retinal implant system, when both the external and internal units of the inductive link are tuned to a same resonant frequency ω_r, the maximal coupling energy can be delivered to the implanted system. In this case, the coil Q factor can be expressed with the following equation:

$$Q_r = \frac{Im(Z_s)}{Re(Z_s)} \approx \frac{\omega_r L_s}{R_s} \tag{9}$$

Combining with equations (1) and (3), equation (9) can be rewritten as

$$Q_r = \frac{2\omega_r N t h}{\rho}\left[\left(\ln\frac{4d}{t}\right)\left(1+\frac{t^2}{24d^2}\cdots\right) - \frac{1}{2} + \frac{43t^2}{288d^2}\cdots\right] \times 10^{-9} \tag{10}$$

Ideally, the Q factor of a coil should be as high as possible in order to minimize the power loss in the device as well as to maximize power transfer efficiency of the system. It can be seen from equation (10) that Q_r can be enhanced by increasing the number of coil layers (N), the coil width (t), and/or the thickness of the conductive layer (h). For an intraocular retinal implant, however, there is not much zoom to improve the coil width due to the small coil dimensions (inner diameter, outer diameter, etc.) confined by the eyeball size. Therefore, a more applicable way to increase a coil's Q factor is to increase the number of stacking layers as well as the thickness of conductive wires.

2.5 Finite element simulation
To validate the effectiveness of the theoretical models, finite element simulations (FES) are performed using a built-in package in CoventorWare (Coventor Inc., Cary, NC). As a demonstration, a coil with two layers of metal is designed, and its electrical characteristics are evaluated using both analytical models and FES, as summarized in Fig. 4 and Table 1.
During the simulation, an octagonal coil is used to approximate a circular shape due to memory constraint in CoventorWare. The coil self-inductance and the ESR are simulated over a frequency range from 10 kHz to 1 GHz. It can be seen that the self-inductance at 1 MHz shows only 2.2% deviation, and the ESR deviates by less than 6%, suggesting good agreement with the analytical models. The slight deviations might be introduced by the approximation of coil shape. Note that L_s and R_s both remain stable at frequencies below 10 MHz, indicating that skin effect or proximity effect is negligible at target frequencies of 1 or 2 MHz.

	OD (mm)	ID (mm)	Trace cross section (μm × μm)	Number of turns	L_s (μH)	R_s (Ω)	C_s (nF)	Q at 1MHz
Calculations	10	3	220 × 2	28 / Layer	5.0	28.9	65.5	1.1
FES	10	3	220 × 2	28 / Layer	4.9	27.4	--	1.12

Table 1. Coil characteristics estimated using both analytical models and FES.

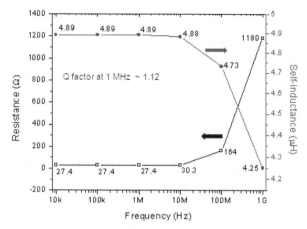

Fig. 4. Simulated self-inductance and ESR of the sample coil.

3. Parylene-metal-Parylene thin film technology

A multi-layer Parylene-metal thin film technology for making the proposed microcoils has been developed (Li et al., 2005). In this approach, thin-film metal conductive wires are sandwiched between multiple layers of Parylene C and interconnections between two adjacent layers are implemented using through holes in the Parylene insulation layer. Fig. 5 depicts a typical process flow for making a dual-metal-layer structure. Briefly, a layer of sacrificial photoresist is optionally spun on a standard silicon wafer, followed by Parylene C deposition (PDS 2120 system, Special Coating Systems, Indianapolis, IN, USA) (*Step 1*). A layer of metal is then deposited on top of the Parylene using an electron beam (e-beam) evaporator (SE600 RAP, CHA Industries, Fremont, CA, USA), and patterned using a wet etching process (*Step 2*). After that, a thin layer of Parylene C is deposited as an insulation layer between two metal layers, and the interconnection vias are selectively opened with oxygen plasma in a reactive ion etch system (RIE) (Semi Group Inc. T1000 TP/CC) using a photoresist mask(*Step 3*). After removing the photoresist mask, the second metal layer is evaporated and patterned, followed by another Parylene C deposition to conformally cover the exposed metal wire (*Step 4*). A photoresist mask is then patterned to expose the contact pads, as well as to define the coil shape (*Step 5*). Finally, oxygen plasma etch is performed to remove unwanted Parylene C, and the entire flexible device is released from the silicon substrate by dissolving the sacrificial photoresist (*Step 6*).

The reliability of the interconnections between nearby metal layers highly depends on the step coverage of the Parylene sidewall during metal evaporation, which can be improved by a slightly isotropic O_2 plasma etch (Meng et al., 2008). A special design of rotating wafer holder inside the e-beam evaporator also helps adjust the angle of attack of metal evaporant for best coverage. Microcoils comprising more than two layers of metal can be fabricated with similar procedure by alternating the Parylene C deposition, interconnection via fabrication, and metal evaporation process steps. Although it is specifically developed for microcoil fabrication, this technology can also be applied to the fabrication of other flexible, implantable devices with multi-layer Parylene-metal structures, such as dual-metal-layer electrode arrays (Rodger et al., 2008).

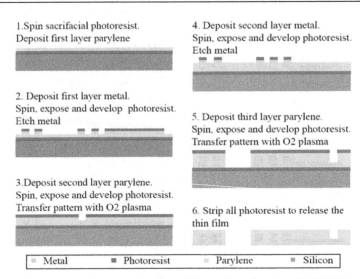

1.Spin sacrifacial photoresist.
Deposit first layer parylene

2. Deposit first layer metal.
Spin, expose and develop photoresist.
Etch metal

3.Deposit second layer parylene.
Spin, expose and develop photoresist.
Transfer pattern with O2 plasma

4. Deposit second layer metal.
Spin, expose and develop photoresist.
Etch metal

5. Deposit third layer parylene.
Spin, expose and develop photoresist.
Transfer pattern with O2 plasma

6. Strip all photoresist to release the
thin film

| Metal | Photoresist | Parylene | Silicon |

Fig. 5. Fabrication process flow of a dual-metal-layer Parylene-based MEMS structure.

For implantable devices, our Parylene-metal thin film technology has several unique advantages compared with conventional semiconductor-based microfabrication technologies. Using biocompatible Parylene directly as the actual substrate greatly simplifies the device integration and packaging procedures. Devices fabricated in this way are very flexible and foldable so that they can be implanted through small surgical incisions, allowing wounds to heal quickly. Moreover, the metal lines are completely padded by the Parylene material, and can therefore withstand repeated bending during surgical handling. Finally, a post-fabrication heat-molding process has been developed to modify the skins into various shapes that match the curvatures of the target implant areas (Tai et al., 2006).

4. Coil designs and fabrication results

4.1 Dual-layered MEMS coil

In this section, a planar coil is designed, which features: 1) dual-layer thin-film metal conductive wires sandwiched between multiple layers of Parylene C, and 2) interconnections between two adjacent layers that are formed by filling the Parylene through holes with PVD metal. Fig. 6 shows the microscope images of a fabricated coil and its interconnection via. This coil has totally 10 turns of wires made of approximately 2000 Å multiple layers of Ti/Au/Ti metallization. Titanium serves as an adhesion promoter to improve the bonding strength between gold and Parylene C. The device has overall dimensions of ~ 9.5 mm in outer diameter, ~ 5 mm in inner diameter, and ~ 11 μm in thickness, limited by the lens capsule size of the human eyes. The interconnection via occupies an area of ~ 0.06 mm^2 with a contact resistance of less than 1 Ω, which can be negligible compared with the total coil ESR. The device is proven to be very flexible and foldable (Fig. 7), easing the procedure of surgical insertion and lessening physical damages in the region of implantation.

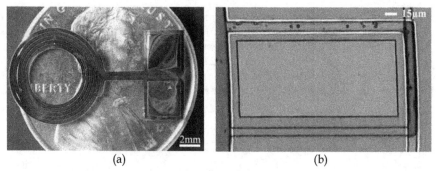

(a) (b)

Fig. 6. (a) A fabricated dual-metal-layer coil sitting on a penny. (b) The microscope image shows the interconnection via between two metal layers. (Li et al., @ 2005 IEEE)

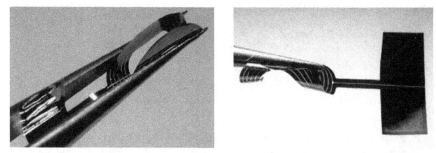

Fig. 7. Demonstration of the coil's flexibility and foldability. (Li et al., @ 2005 IEEE)

The electrical properties of the fabricated coil are characterized experimentally. Recall equation (7) in Section 2, by setting the derivation of the real part to zero and equating the imaginary part to zero, the self-inductance (L_s) and the parasitic capacitance (C_s) can be extracted using equations (11) and (12), where ω_0 is defined as the frequency at which the real part of the impedance is maximum, and ω_z is the zero-reactance frequency at which the imaginary part of the impedance is zero (Wu, 2003).

$$L_S = \frac{R_s}{\sqrt{2(\omega_0^2 - \omega_z^2)}}, \tag{11}$$

$$C_S = \frac{\sqrt{2(\omega_0^2 - \omega_z^2)}}{R_s(2\omega_0^2 - \omega_z^2)}. \tag{12}$$

For the coil in Fig. 6, the ESR (R_s) is measured to be around 72 Ω and the resistivity of e-beam deposited gold is calculated to be around 2.25×10^{-6} $\Omega \cdot$cm. This number agrees with the resistivity of bulk gold (2.2×10^{-6} $\Omega \cdot$cm), implying that the E-beam evaporated metal is void-free. The coil impedance is swept with an HP 4192A LF impedance analyzer over a frequency range from 5 Hz to 13 MHz. From the impedance versus frequency curves (Fig. 8), f_0 and f_z can be read with values of 7.5 MHz and 3.3 MHz respectively. Knowing R_s, ω_0, and ω_z, the coil self-inductance and capacitance are therefore calculated as $L_s = 1.19$ μH and

C_s = 201 pF. The theoretical numbers are also calculated using the abovementioned equations, and the fitting curves are plotted in Fig. 8, in comparison with the measured curves. The experimental data matches the theoretical calculations closely, with deviations of less than 19%. These errors may be attributed to the simplification of the 3-element model as well as interferences from the measurement instruments. The Q factor of the coil is obtained to be approximately 0.1 at the target frequency of 1 MHz, as expected from the design.

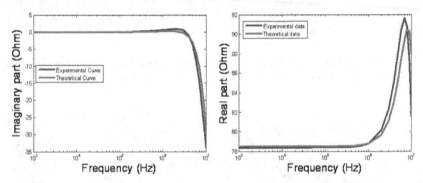

Fig. 8. Impedance measurement and curve fitting using the 3-element model: (a) Imaginary part; (b) Real part. (Red curves correspond to theoretical parameters of the fabricated coil: L_s = 1.0 μH, R_s = 67 Ω and C_s = 183 pF.)

The data and power transfer performances have also been verified using a custom designed data link at the University of California, Santa Cruz (UCSC). The testing waveforms are shown in Fig. 9, where the blue curve represents the data driving signal on the primary stage, the green curve represents the voltage across the primary coil, and the purple curve represents the receiving voltage across the secondary coil.

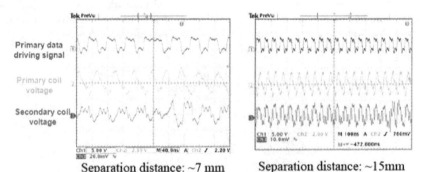

Fig. 9. Inductive coupling test waveforms: (a) received signal is 25 mV peak to peak; (b) received signal is 15 mV peak to peak.

While successful data transmission through our coils has been demonstrated, it is noted that this device has no driving capability due to its small Q factor (~ 0.1), meaning that the power cannot be delivered to the load. Therefore, enhancing coil's Q factor to achieve a higher power transfer efficiency is crucial for designing the next generation of coils.

Experiments have been done at UCSC to study the feasibility of using MEMS coils as the receiver coil for the current inductive link design. A rough estimate is that, in the worst case, a minimal Q factor of 10 will be needed in order to deliver ~100 mW for chip operation and stimulation. From the theoretical analysis herein, it is known that the Q factor can be enhanced by increasing the metal thickness and/or the number of metal layers. However, e-beam evaporated metals are usually limited in film thickness due to high process cost. Electroplated and sputtered metals can be thicker alternatives but their qualities, such as density and conductivity, are typically not as good as evaporated metals. This problem becomes more serious especially when devices are implanted inside harsh biological environments. From the device design aspect, increasing metal layers is more practical for the Q factor enhancement of MEMS coils, thus a fold-and-bond technology emerges as a good candidate.

4.2 Fold-and-bond coil
In the concept of fold-and-bond technology (Fig. 10), two or more thin-film planar spiral coil segments are fabricated from the same batch so that each segment has identical self-inductance and ESR, denoted by L_s and R_s, respectively. A new coil can then be formed by stacking n segments together in either parallel or series connections. For parallel stacking, particularly, the new coil will have identical inductance but n-times larger in equivalent metal thickness, resulting in an n-times lower ESR. As for series stacking, the resistance remains the same while the inductance increases by n-times because of the mutual inductance between adjacent layers. According to the definition of the Q factor, both approaches can achieve an n-times Q factor enhancement. In this section, the series stacking configuration is used to demonstrate the technology concept.

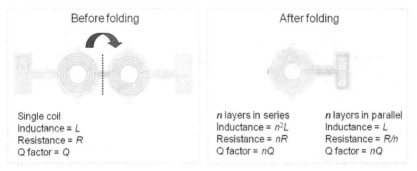

Fig. 10. Concept of the fold-and-bond technology for Q factor enhancement. A coil with one fold is depicted for representation.

Fold-and-bond coil's fabrication involves the dual-metal-layer Parylene/metal skin realization and a post-fabrication thermal bonding process (Li et al., 2008). The Parylene-metal skin with two buried layers of metal is first fabricated in the same manner as described in Fig. 5, in which one layer of metal is used to form the conductive wires of the coil, while the other layers is used to make the interconnections between the layers. This thin film skin can be folded and stacked into multiple layers because of the flexibility of Parylene C. While hand alignment under an optical microscope is used at the current stage, special alignment jigs can be custom designed in the future to achieve precise alignment of different layers. During the thermal bonding procedure, the folded device is sandwiched between two glass slides covered with aluminum sheets, which can avoid Parylene sticking on the glass. The whole unit is placed in a vacuum oven and bonded at 230 °C for two days. External pressure can be applied as needed to enable Parylene-Parylene bonding at moderate temperatures. The vacuum pressure is controlled at ~ 10 Torr to prevent Parylene C from unwanted oxidation in air at an elevated temperature.

Two Parylene-based skins with dual-layer embedded metal have been fabricated, as shown in Fig. 11. These prototypes are specifically designed for intraocular retinal prosthesis with the design parameters described in Table 2. The thickness of metal wires is increased to ~ 2 µm in order to further reduce the coil's DC resistance. The metal is covered with ~ 3.4 µm Parylene C on each side with the lead contact vias open. Fig. 12 shows the final devices after folding and thermal bonding. Misalignments of 10 µm to 30 µm have been observed, which is due to the lack of control with the hand alignment.

Type	Total number of turns	R_s (Ω)	L_s (μH)	C_s (pF)	Q at 1MHz
I	28	29	2.9	65.5	1.1
II	48	90	14.8	64.5	1

Table 2. Design specifications of two fold-and-bond coils.

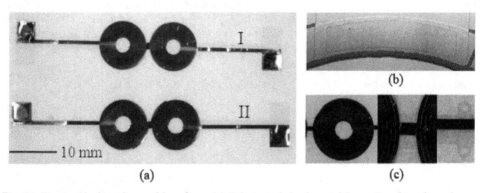

(a)   (c)

Fig. 11. Devices before thermal bonding: (a) Fabricated dual-metal-layer Parylene-based skins; (b) Microscope image of an interconnection via between two metal layers; (c) Photos of device details (from left to right): conductive wires of the coil, folding junction and suturing holes.

The devices still remain flexible after bonding (Fig. 13 (a)), indicating that annealing at a temperature below the melting point of Parylene C (290 °C) (Harder et al. 2002) will not alter the mechanical flexibility of the material. The DC resistances of the samples are measured before and after bonding with no significant change observed (Table 3), confirming the ductility and durability of metal traces. Stretching marks and Parylene cracks are found along the folding line after thermal treatment, which is caused by stress concentration during folding. Additional Parylene coating can be performed after thermal bonding to conformally cover these cracks in order to ensure a good sealing for final devices.

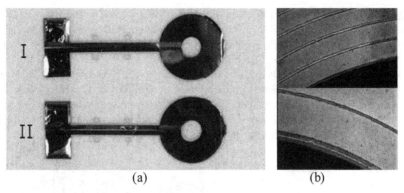

(a)   (b)

Fig. 12. Devices after thermal bonding: (a) Fold-and-bond coils after thermal bonding. (b) Overlapping metal wires with misalignments of 10 μm to 30 μm.

(a)   (b)

Fig. 13. (a) Demonstration of device's flexibility; (b) Stretching marks along the folding line.

The electrical characteristics of the fabricated fold-and-bond coils are studied before and after thermal bonding. The lump parameters are extracted from the same 3-element model, as summarized in Table 3. As expected, the self-inductance and Q-factor of the coil are increased by more than 90% for both coil designs after folding. Changing the wire width has no significant impact on the parasitic capacitance, indicating the capacitance between adjacent layers is dominant over the capacitance between turns. Coil prototype I shows a much higher ESR compared to its theoretical value (29 Ω), which may be attributed to the non-uniformity of metal thickness. Overall, the measured values show good agreement with the theory predictions, demonstrating that the theoretical model can predict the coil properties effectively.

	Type	R_s (Ω)	L_s (μH)	C_s (pF)	Calculated Q at 1MHz	Q Increase
I	Before	41.2	2.9	--	0.44	--
	After	41.8	5.7	62.1	0.85	95%
II	Before	91.3	8.6	--	0.58	--
	After	92.2	16.3	70.8	1.11	91%

Table 3. Measured electrical parameters of fold-and-bond coils using the 3-element model.

To study the power transfer efficacy, the fabricated devices are tested using a simplified inductive link, as shown in Fig. 14. In this setup, the transmitter coil is hand-wound with a self-inductance of ~ 23 μH and a series resistance of ~ 1.5 Ω. The inner diameter of the transmitter coil is optimized to be ~ 30 mm (Ko et al., 1977). The receiver coil is the fold-and-bond coil presented in Fig. 12. During the measurements, the primary stage is driven by an HP E3630A function generator with a sinusoidal input signal of 20 V peak to peak. Both the primary and secondary circuits are subject to parallel resonance with the same resonant frequencies of ~ 1 MHz. To minimize environmental interferences, both the transmitter and receiver coils are covered with aluminum foils as electromagnetic shielding.

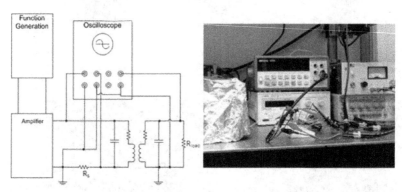

Fig. 14. Experimental setup of power transmission measurement and its circuit diagram.

Preliminary experiments have been performed. A 10 Ω series resistor (R_s) is incorporated in the primary stage to monitor the output current from the amplifier. The voltage across the transmitter coil is then calculated by subtracting the resistor voltage from the output voltage of the amplifier. The transferred power, which is defined as the power delivered to a 1 kΩ load resistor (R_{load}), can be obtained by directly measuring the voltage across the load resistor. The power transfer efficiency of two coil prototypes, which is calculated as the ratio of the transferred power to the total output power from the amplifier, is plotted in Fig. 15, as

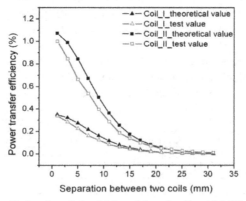

Fig. 15. Power transfer efficiencies of the fold-and-bond coils at 1 MHz, as functions of separation distances of the coil pair.

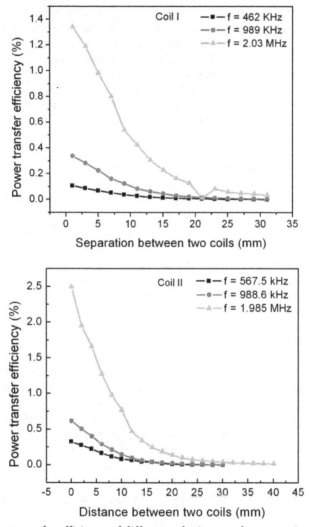

Fig. 16. The power transfer efficiency of difference devices vs. the separation distance of the coil pair.

functions of the separation distance between the coil pairs. For comparisons, theoretical data are also studied based on the models discussed elsewhere (Ko et al., 1977), which match the testing results within a reasonable range. As expected, the coil prototype II with higher self-inductance and Q factor exhibits higher power transfer efficiency at the same separation distances. The power transfer efficiencies at different operation frequencies have also been investigated, as plotted in Fig.16. The results show that the power transfer efficiency can be enhanced by more than 3 times when the operation frequency is changed from ~1 MHz to ~2 MHz. This can be attributed to the effective Q factor increases by almost two times as the frequency goes up.

Nevertheless, the power transfer efficiency of the current coil prototypes is still below 0.4% at the target implantation depth of ~ 15 mm, which is insufficient to drive electronics for high-density retinal stimulation. Whereas more power can be generated from the primary stage to compensate the low efficiency, the overheating issue of the coils becomes prominent. Therefore, future direction for coil optimization will mainly focus on increasing the number of metal layers to improve the Q factor. Theoretical evaluation predicts that at least 20 layers of metal will be required in order to achieve a reasonable Q factor of around 10, with a given gold thickness of ~ 2 µm.

4.3 Challenges in packaging

Device packaging for retinal prostheses presents several challenges. Electrical and fluidic isolation must be maintained to avoid device failures upon exposure to corrosive eye fluid. To ensure long-term use of implants, the interface between device and biological tissues should be stable and biocompatible in order to minimize inflammation and immune response. Packaging complication is also one of the significant challenges for high-density neural simulation/recording. Careful selection and evaluation of packaging materials and tools are necessary to address such challenges. We have studied the packaging performance of Parylene C using accelerated lifetime soak testing in heated saline. Preliminary results estimate that the lifetime of Parylene-coated metal at body temperature (37 °C) is more than 60 years, suggesting good packaging performance of Parylene C. Experimental details on Parylene packaging evaluation are discussed elsewhere (Li et al., 2010).

A chip-level integrated interconnect (CL-I²) packaging method has also been explored for integration of microcoils with CMOS integrated circuit (IC) chips and high-density prosthetic electrodes (Rodger et al., 2005 & Li et al. 2010). In this method, all the IC components necessary for a retinal implant can be embedded in silicon cavities and functional Parylene-based MEMS devices (e.g. microcoil and high-density electrode array) can then be fabricated on the same platform using the abovementioned Parylene-metal thin film technology. Chip-to-microdevice interconnections can be constructed using standard microfabrication techniques such as photolithography and metal etching; therefore, eliminating wire-bonding, bump-bonding, or soldering steps. Whereas initial experiments show promising results, continuous investigations will be necessary to collect more data in order to optimize integration process and to further refine our knowledge of Parylene packaging behavior.

5. Conclusion

In this work, various types of MEMS coils have been designed and fabricated using the Parylene-metal-Parylene skin technology. Experiments have been performed to measure the electrical properties of the coils and the results show good agreement with the theoretical values. The data transfer effect has been successfully demonstrated with the telemetry link setup. However, the power transfer efficiency at the separation distance of 15 mm is below 0.4%, which is relatively low for high-density retinal stimulation. Given the constraints of device geometries, it is believed that increasing the number of metal layers will be the most effective and applicable way to enhance the Q factor of microcoils. According to the analytical models, at least 20 layers of metal will be needed in order to achieve a coil Q factor of approximately 10. While direct fabrication could be too complicated to be carried out, the fold-and-bond technology has proven itself as a very promising fabrication technology. Although specifically tailored to the needs of retinal prostheses, because our coils are fully micromachined in a way compatible with multielectrode arrays and the Parylene-based embedded chip packages, these devices can be easily integrated with various system components to achieve a new range of true system solutions for both biomedical and non-biomedical applications.

6. Acknowledgment

This work is supported in part by the Engineering Research Center Program of the National Science Foundation under Award Number EEC-0310723 and by a fellowship from the Whitaker Foundation (D.R.). The authors would like to acknowledge Dr. Yang Zhi for his supports on the coil testing. We also want to thank Mr. Trevor Roper, Dr. Wen-Cheng Kuo, and other members at the Caltech Micromachining Laboratory for assistance with device simulation and fabrication.

7. References

Artificial Retina Project. (2007). Retinal Diseases: Age-Related Macular Degeneration and Retinitis Pigmentosa. Available from
http://artificialretina.energy.gov/diseases.shtml.
Bennett, J.; Tanabe, T.; D. Zeng, Sun, Y.; Kjeldbye, H.; Gouras, P. & Maguire, A.M. (1996). Photoreceptor Cell Rescue in Retinal Degeneration (RD) Mice by In Vivo Gene Therapy. *Nature Medicine*, Vol. 2, (June 1996), pp. 649-654, ISSN 1078-8956.
Chen, P.J.; Kuo, W.C.; Li, W. & Tai, Y.C. (2006) Q-enhanced Fold-and-bond MEMS Inductors. *Proc. IEEE Int. Conf. on Nano/Micro Engineered and Molecular Systems*, ISBN 978-1-4244-1907-4, Sanya, Hainan Island, China, January 2008.
Chow, A.Y.; Chow, V.Y.; Packo, K.H.; Pollack, J.S.; Peyman, G.A. & Schuchard, R. (2006). The Artificial Silicon Retinamicrochip for the Treatment of Vision Loss from Retinitis Pigmentosa. *Arch Ophthalmol*, Vol. 122, (April 2004), pp. 460-469, ISSN 0003-9950.
Dwight, H.B. (1945). *Electrical Coils and Conductors*, McGraw-Hill, ASIN B00071T6Z0, New York, United States.
Harder, T.; Yao, T.J.; He, Q.; Shih, C. Y. & Tai, Y.C. (2002). Residual Stress in Thin-film Parylene C. *Proc. IEEE Int. Conf. on Micro Electro Mechanical Systems*, ISBN 0-7803-7185-2, Las Vegas, United States, January 2002.
Horch K.W. & Dhillon, G.S. (2004). *Neuroprosthetics Theory and Practice (Series in Bioengineering & Biomedical Engineering-Vol.2)*, World Scientific Publishing Company, ISBN 9812380221, Singapore.
Humayun, M.S.; Propst, R.; Eugene de Juan Jr.; McCormick, K. & Hickingbotham, D. (1994). Bipolar Surface Electrical Stimulation of the Vertebrate Retina. *Arch Ophthalmol*, Vol. 112, (January 1994), pp. 110-116, ISSN 0003-9950.

Humayun, M.S.; Eugene de Juan Jr.; Weiland, J.D.; Dagnelie, G.; Katona, S.; Greenberg, R. & Suzuki, S. (1999). Pattern Electrical Stimulation of the Human Retina. *Vision Research*, Vol. 39, (July 1999), pp. 2569-2576, ISSN 0042-6989.

Javaheri, M.; Hahn, D.S.; Lakhanpal, R.R.; Weiland, J.D. & Humayun, M.S. (2006). Retinal Prostheses for the Blind. *Ann Acad Med*, Vol. 35, (March 2006), pp. 137-144, 2006, ISSN 0304-4602.

Kim, S.Y.; Sadda, S.; Pearlman, J.; Humayun, M.S.; Eugene de Juan Jr. & Green, W.R. (2002). Morphometric Analysis of the Macula in Eyes with Disciform Age-realted Macular Degeneration. *Retina*, Vol. 22, No. 4, (August 2002), pp. 471-477, ISSN 0275-004X.

Ko, W.H.; Liang, S.P. & Fung, C.D.F. (1977). Design of Radio-frequency Powered Coils for Implant Instruments. *Med. Bio. Eng. Comput.*, Vol. 15, (November 1977), pp. 634-640, ISSN 0140-0118.

Li, W.; Rodger, D.C.; Weiland, J.D; Humayun, M.S. & Tai, Y.C. (2005). Integrated Flexible Ocular Coil for Power and Data Transfer in Retinal Prostheses. *Proc.27th Ann. Int. IEEE-EMBS Conf.*, ISBN 0-7803-8741-4, Shanghai, China, January 2005.

Li, W.; Rodger, D.C.; Meng, E.; Weiland, J.D.; Humayun, M.S. & Tai, Y.C. (2006). Flexible Parylene Packaged Intraocular Coil for Retinal Prostheses. *Proc. 4th Int. IEEE-EMBS Special Topic Conf. on Microtechnologies in Medicine and Biology*, ISBN 1-4244-0338-3, Okinawa, Japan, May 2006.

Li, W.; Rodger, D.C. & Tai, Y.C. (2008). Implantable RF-coiled Chip Packaging. *Proc. IEEE Int. Conf. on Micro Electro Mechanical System*, ISSN 1084-6999, Tucson, United States, January 2008.

Li, W.; Rodger, D. C.; Meng, E.; Weiland, J. D.; Humayun, M. S. & Tai, Y.-C. (2010) Wafer-Level Parylene Packaging with Integrated RF Electronics for Wireless Retinal Prostheses. *Microelectromechanical Systems, Journal of*, Vol. 19, (June 2010) pp. 735-742, ISSN 1057-7157.

Licari J. J. & Hughes, L. A. (1990). *Handbook of Polymer Coating for Electronics: Chemistry, Technology, and Applications*, William Andrew Publishing/Noyes, ISBN 081551235X, Park Ridge, New Jersey, United States.

MacLaren, R.E.; Pearson, R.A.; MacNeil, A.; Douglas, R.H.; Salt, T.E.; Akimoto, M.; Swaroop, A.; Sowden, J.C. & Ali, R.R. (2006). Retinal Repair by Transplantation of Photoreceptor Precursors. *Nature*, Vol. 444, (November 2006), pp. 203-207, ISSN 0028-0836.

Meng E., Li P.Y. & Tai, Y.C. (2008) Plasma Removal of Parylene C, *Journal of Micromechanics and Microengineering*, Vol 18, (February 2008), pp. 045004, ISSN 1361-6439.

Mokwa, W.; Goertz, M. C.; Krisch, K. I.; Trieu, H.-K. & Walter, P. (2008) Intraocular Epiretinal Prosthesis to Restore Vision in Blind Humans. *Proc. 30th Ann. Int. IEEE EMBS Conf.*, ISBN 978-1-4244-1814-5, Vancouver, British Columbia, Canada, August 2008.

Norton, E.W.D.; Marmor, M.F.; Clowes, D.D.; Gamel, J.W.; Barr, C.C.; Fielder, A.R.; Marshall, J.; Berson, E.L.; Rosner, B.; Sandberg, M.A.; Hayes, K.C.; Nicholson, B.W.; Weigel-DiFranco, C.; Willett, W.; Felix, J.S. & Laties, A.M. (1993). A Randomized

Trial of Vitamin A and Vitamin E Supplementation for Retinitis Pigmentosa. *Arch Ophthalmol*, Vol. 11, (June 1993), pp. 1460-1466, ISSN 0003-9950.

Rizzo, J.F.; Wyatt, J.L.; Loewenstein, J.; Montezuma, S.; Shire, D.B.; Theogarajan, L. & Kelly, S.K. (2004). Development of a Wireless, Ab Externo Retinal Prosthesis. *Invest Ophthalmol Vis. Sci.*, Vol. 45, pp. 3399.

Rizzo, J.F. (2011). Update on Retinal Prosthetic Research: The Boston Retinal Implant Project. *Journal of Neuro-Ophthalmology*, Vol.32, (June 2011), pp. 160-168, ISSN 1070-8022.

Rodger, D. C.; Weiland, J. D.; Humayun, M. S. & Tai, Y.-C. (2005). Scalable Flexible Chip-level Parylene Package for High Lead Count Retinal Prostheses. *Proceedings of the 13th International Conference on Solid-State Sensors, Actuators and Microsystems*, ISBN 0-7803-8994-8, Seoul, South Korea, June 2005.

Rodger, D.C.; Weiland, J.D.; Humayun, M.S. & Tai, Y.C. (2006), Scalable High Lead-count Parylene Package for Retinal Prostheses. *Sensors and Actuators B: Chemical*, Vol. 117, (September 2006), pp. 107-114, ISSN 0925- 4005.

Rodger, D.C.; Fong, A.J.; Li, W.; Ameri, H.; Ahuja, A.K.; Gutierrez, C.; Lavrov, I.; Zhong, H.; Menon, P.R.; Meng, E.; Burdick, J.W.; Roy, R.R.; Edgerton, V.R.; Weiland, J.D.; Humayun, M.S. & Tai, Y.C. (2008). Flexible Parylene-based Multielectrode Array Technology for High-density Neural Stimulation and Recording. *Sensors and Actuators B: Chemical*, Vol. 132, (June 2008), pp. 449-460, ISSN 0925- 4005.

Stieglitz, T.; Haberer, W.; Lau, C. & Goertz, M. (2004). Development of an Inductively Coupled Epiretinal Vision Prosthesis. *Proc. Int. IEEE Eng. in Med. and Biol. Soc. Meet.*, ISBN 0-7803-8439-3, San Francisco, CA, USA, September 2004.

Tai, Y.C.; Rodger, D.C.; Li, W. & Tooker, A. (2006). *Method for Decreasing Chemical Diffusion in Parylene and Trapping at Parylene-to-parylene Interfaces*, US Patent Applicatio 11/408809. Available from
http://www.freepatentsonline.com/y2006/0255293.html

Tropepe, V.; Coles, B.L.K.; Chiasson, B.J.; Horsford, D.J.; Elia, A.J.; McInnes, R.R. & Kooy, D.V.D. (2000). Retinal Stem Cells in the Adult Mammalian Eye. *Science*, Vol. 287, (March 2000), pp. 2032-2036, ISSN 1934-7391.

Wong, Y.T.; Chen, S.C.; Seo, J.M.; Morley, J.W.; Lovell, N.H. & Suaning, G.J. (2009). Focal Activation of the Feline Retina via a Suprachoroidal Electrode Array. *Vision Research*, Vol 49, (May 2009), pp. 825-833, ISSN 0042-6989.

Weiland J.D. & Humayun, M.S. (2008). Visual Prosthesis. *Proceedings of the IEEE*, Vol. 96, (July 2008), pp. 1076-1084, ISSN 0018-9219.

World Health Organization. (2011). Prevention of Blindness and Visual Impairment. Available from http://www.who.int/blindness/causes/priority/en/index.html.

Wu, C.; Tang, C. & Liu, S. (2003). Analysis of On-chip Spiral Inductors Using the Distributed Capacitance Model. *IEEE J. of Solid-State Circuits*, Vol. 38, (June 2003), pp. 1040-1044, ISSN 0018-9200.

Wu J. (2003). *Inductive Links with Integrated Receiving Coils for MEMS and Implantable Applications* PhD thesis, University of Notre Dame. Available from
http://etd.nd.edu/ETD-db/theses/available/etd-09302003-162720/.

Zrenner, E.; Stett, A.; Weiss, S.; Aramant, R.B.; Guenther, E.; Kohler, K.; Miliczek, K.-D.; Seiler, M.J. & Haemmerle, H. (1999). Can Subretinal Microphotodiodes Successfully Replace Degenerated Photoreceptors? *Vision Research*, Vol. 39, (July 1999), pp. 2555-2567, ISSN 0042-6989.

Zolfaghari, A. Chan, A. & Razavi, B., Stacked Inductors and Transformers in CMOS Technology. *IEEE J. of Solid-State Circuits*, Vol. 36, (April 2001), pp. 620-628, ISSN 0018-9200.

3

Acoustic Wave Based MEMS Devices, Development and Applications

Ioana Voiculescu and Anis N. Nordin
¹City College of New York
²International Islamic University
¹USA
²Malaysia

1. Introduction

Acoustic waves based MEMS devices offer a promising technology platform for a wide range of applications due to their high sensitivity and the capability to operate wirelessly. These devices utilize an acoustic wave propagating through or on the surface of a piezoelectric material, as its sensing mechanism. Any variations to the characteristics of the propagation path affect the velocity or amplitude of the wave.

Important application for acoustic wave devices as sensors include torque and tire pressure sensors (Cullen et al., 1980; Cullen et al., 1975; Pohl et al., 1997), gas sensors (Levit et al., 2002; Nakamoto et al., 1996; Staples, 1999; Wohltjen et al., 1979), biosensors for medical applications (Andle et al., 1995; Ballantine et al., 1996; Cavic et al., 1999; Janshoff et al., 2000), and industrial and commercial applications (vapor, humidity, temperature, and mass sensors) (Bowers et al., 1991; Cheeke et al., 1996; Smith, 2001; N. J. Vellekoop et al., 1999; Vetelino et al., 1996; Weld et al., 1999).

This chapter is focused on two important applications of the acoustic-wave based MEMS devices; (1) biosensors and (2) telecommunications. The technological advancement of the micro-electromechanical systems (MEMS) facilitated the development of biosensors and various devices for telecommunications.

There has been increasing interest to develop miniature, portable and low-cost biosensors fabricated using MEMS technologies. For biological applications the acoustic wave device is integrated in a microfluidic system and the sensing area is coated with a biospecific layer. When a bioanalyte interacts with this sensing layer, physical, chemical, and/or biochemical changes are produced. Typically, mass and viscosity changes of the biospecific layer can be detected by analyzing changes in the acoustic wave properties such as velocity, attenuation and resonant frequency of the sensor. An important advantage of the acoustic wave biosensors is simple electronic readout that characterizes these sensors. The measurement of the resonant frequency or time delay can be performed with high degree of precision using conventional electronics.

Currently, a limitation of acoustic wave devices for biological applications is that they reuire expensive electronic detection systems, such as network analyzers. A final product aimed at the end user market must be small, portable and packaged into a highly integrated cost effective system. For acoustic wave biosensors integrated in a lab-on-chip device, sample pre-treatment, purification and concentration, as well as a good interface between the user and the integrated sensing system also need to be developed in the future

Historically, acoustic wave devices are widely used in telecommunications industry, primarily in mobile cell phones and base stations. Surface Acoustic Wave (SAW) devices are capable of performing powerful signal processing and have been successfully functioning as filters, resonators and duplexers for the past 60 years. Although SAW devices are technological mature and have served the telecommunication industry for several decades, these devices are typically fabricated on piezoelectric substrates and are packaged as discrete components. The wide flexibility and capabilities of the SAW device to form filters, resonators there has been the motivation to integrate such devices on silicon substrates (Nordin et al., 2007; M. J. Vellekoop et al., 1987; Visser et al., 1989). Standard Complementary Metal Oxide Semiconductor (CMOS) technology with additional MEMS post-processing was used for the fabrication of a CMOS SAW resonator in 0.6 µm AMIs CMOS technology (Nordin et al., 2007). The advantage of using standard CMOS technology for the fabrication of a SAW resonator is that active circuitry can be fabricated adjacent to the CMOS resonator on the same electronic chip.

Telecommunication devices based on acoustic waves have different requirements compared to biosensors. The biosensors operates at frequencies in the range of MHz where acoustic wave devices operating as a filter or resonator are expected to operate at high frequencies (GHz) and have high quality factors and low insertion losses. With the advancement in lithographic techniques, the acoustic wave based devices have the advantage of meeting the stringent requirement of telecommunication industry of having Qs in the 10,000 range and silicon compatibility.

A simple, robust, cheap packaging method is also critical for the commercialization of the acoustic wave devices. The integration of the acoustic wave based MEMS biosensor in the microfluidic system is a complex matter. The integration technique is influenced by the sensor fabrication process and the type of the biological applications. In some applications the sensor could be embedded in a microfluidic reservoir. In the case when the biological application requires different biological solutions to be introduced on the sensor sensitive area the biosensor could be embedded in a microfluidic channel. The packaging of the acoustic wave devices used for telecommunication is less complicated since these devices are embedded in the package and do not need to be in contact with liquid.

2. Acoustic wave MEMS devices as biosensors

There has been increasing interest to develop miniature, portable and low-cost biosensors fabricated using MEMS technologies. MEMS technology has been adopted from the integrated circuit (IC) industry and applied to the miniaturization of a large range of systems including acoustic wave based devices. Recent technological advancement of MEMS processes allows the fabrication of thin piezoelectric films and the integration of acoustic wave based devices, and electronics on a common silicon substrate. The acoustic wave MEMS biosensors presented in this chapter could be categorized in two main groups; resonators and delay lines.

For biological applications the Acoustic Wave Based MEMS devices are integrated in a microfluidic system and the sensing area is coated with a biospecific layer. When a bioanalyte interacts with this sensing layer, physical, chemical, and/or biochemical changes are produced. Typically, mass and viscosity changes of the biospecific layer can be detected by analyzing changes in the acoustic wave properties such as velocity, attenuation and resonant frequency of the sensor. An important advantage of the acoustic wave biosensors is simple electronic readout that characterizes these sensors. The measurement of the resonant frequency or time delay can be performed with high degree of precision using conventional electronics.

The Sauerbrey equation correlates the changes of the resonant frequency of an acoustic wave resonator with the mass deposited on it. The acoustic wave propagating on a piezoelectric substrate is generated and received using IDTs. In the case of a biosensor resonator, the cell to be analyzed or the antibody layer for protein marker detection are added on the IDTs. This will cause a shift of the resonant frequency due to the increasing of mass, where f_i and f_o are the resonant frequencies before and after loading the sensor.

The Sauerbrey equation is defined as;

$$\Delta f = -\frac{2f_0^2 \Delta m}{A\sqrt{\rho_q \mu_q}} = -2.26 \cdot 10^6 f_0^2 \frac{\Delta m}{A} \tag{1}$$

where $$f_o - f_i = \Delta f \tag{2}$$

From (1) the change Δf of the resonant frequency of the piezoelectric crystal is directly proportional to the mass loaded on the acoustic wave resonator, where Δm is expressed in g and Δf and f_0 in Hz (Skládal, 2003).

Generally, the acoustic wave MEMS resonators employed for biosensing applications are FBAR and acoustic wave based delay lines. For FBAR type biosensor the excitation electrodes are fabricated at both sides of the piezoelectric substrate and the acoustic waves propagate through the volume of the substrate. The detection mechanisms occur at the opposite surfaces of the piezoelectric substrate.

The acoustic wave based delay lines reported in the literature as MEMS biosensors are surface acoustic wave (SAW) delay lines that consists of two sets of interdigitated transducers (IDT)s fabricated on the same side of a thin layer of piezoelectric material. The acoustic wave is produced by one set of IDTs and the second set of IDTs is used to detect the acoustic wave. In the case of a biosensor, the surface between these two sets of IDTs is covered with a biological layer sensitive to the analyte to be detected, as illustrated in Fig. 1.

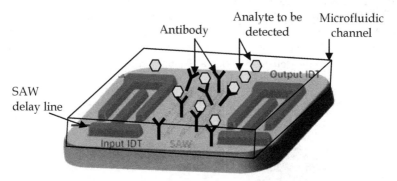

Fig. 1. SAW delay line biosensor integrated in a microfluidic channel. The surface between the IDTs is coated with antibodies sensitive to the analyte to be detected. The analyte

molecules binding to the immobilized antibodies on the sensor surface influence the velocity of the SAW and hence the output signal generated by the driving electronics.

The absorption of the analyte on the sensitive layer will produce a time delay in the acoustic wave propagation. The main disadvantage of the acoustic wave based devices when used as biosensors is the degradation of performance due to liquid damping. In liquid the quality factor Q drops (usually more than 90% reduction) and negatively affects the device sensitivity. Since most of the biological applications are performed in liquid only few types of acoustic wave devices could be integrated in microfluidic channels, without significant degradation of the sensor performance.

2.1 Film Bulk Acoustic wave Resonators (FBAR)
In recent years the thin film fabrication technology has made substantial progress particularly in view of high frequency resonators. In the case of MEMS-based FBAR resonators the expensive single crystalline substrates used for quartz crystal microbalance (QCM) resonators, could be replaced with a large range of thin piezoelectric films. FBAR resonators that are fabricated from a thin piezoelectric substrate and the excitation electrodes are fabricated at both sides of the piezoelectric substrate. The FBAR resonators could be integrated in a microfluidic system and successfully used for biosensing applications because of low damping of the acoustic wave in the liquid (Gabl et al., 2003; Weber et al., 2006; Wingqvist et al., 2005).

A MEMS FBAR biosensor with aluminum nitride (AlN) as piezoelectric film is illustrated in Fig. 2 (Wingqvist et al., 2005). The thickness of the AlN film is 2μm. Bottom and top Al electrodes were patterned with standard lithography and etching processes. The overlap of the top and bottom electrode defines the active area where the acoustic wave is generated. In order to fabricate a shear acoustic wave FBAR, the AlN thin films were grown with the crystallographic axis inclined with an angle of 30° relative to the surface normal. The shear mode is preferred for liquid application instead of the longitudinal mode because of low-loss operation in liquid and small reduction of the quality factor Q. The silicon wafer was etched from the back side to fabricate a free standing membrane used to isolate the resonator acoustically from the substrate and define a cavity. This cavity was further connected to a microfluidic transport system for analyte delivery to the bottom electrode of the resonator. For this type of FBAR, the bottom electrode is the sensing electrode. An Au layer was thermally evaporated onto the bottom Al electrode to create a biochemically suitable surface for subsequent tests. The sensor was tested with different concentration of albumin in solution and the detection limit was 0.3 ng/cm².

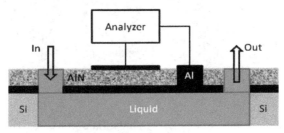

Fig. 2. Schematic of a shear wave FBAR with a microfluidic transport system.

Another FBAR resonator used for biological applications, illustrated in Fig. 3, is fabricated from a thin film of ZnO (Zhang et al., 2009). This FBAR resonator was fabricated from a <100> silicon wafer. A thin silicon dioxide layer was thermally grown on the wafer, followed by the low pressure chemical vapor deposition (LPCVD) of a thin SiN layer. An Al film was evaporated and patterned on the SiN layer as the bottom electrode and determines the effective area of the FBAR. The ZnO piezoelectric layer with the thicknesses from 0.55 μm to 4.2 μm was sputtered on the Al bottom electrode. Layers of Cr/Au representing the top electrode were sputtered and patterned by lift-off over the ZnO layer. Au was chosen as the top electrode due to its excellent conductivity and good affinity for biomolecular binding. The microfluidic channel was fabricated from a 3 μm thick parylene film deposited on the top electrode. The silicon wafer was backside etched using deep reactive ion etch (DRIE) to release the SiN membrane. The SiO$_2$ layer underneath the SiN layer was removed by wet etching, see Fig. 3a. The top Au electrode of the FBAR represents the sensing electrode and in the case of biological applications will be coated with a thin layer of biological material. This sensor is demonstrated a quality factor Q in liquid of 120. Q is improved by integrating a microfluidic channel on FBAR, which confines the liquid to a height comparable to the acoustic wavelength. However, this FBAR biosensor is sensitive to temperature variations. Increasing the temperature degrades Q, resulting in degradation of resolution.

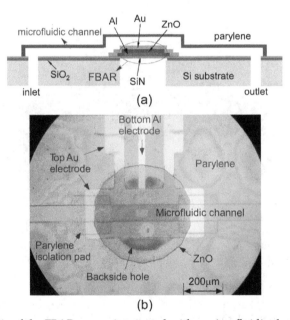

Fig. 3. (a) Schematic of the FBAR sensor integrated with a microfluidic channel and
(b) Top view of the fabricated FBAR sensor; a microfluidic channel run across the FBAR sensor (Zhang et al., 2009). (© [2009] IEEE.) Used with permission.

Typical MEMS FBAR's have high quality factor Q in the air and vacuum which range from a few hundreds to a few thousands. High Qs are possible since the acoustic wave generated in the FBAR is well confined by the very large acoustic impedance mismatch between the solid materials and air. However, for biological applications the FBAR resonators are immersed in liquid and the solid-liquid interface becomes leaky for acoustic waves since the impedance mismatch is small, resulting in the degradation of Q of FBAR. To improve Q of FBAR in liquid, it is recommended that the microfluidic channels on top of the FBAR have the heights comparable to the acoustic wavelength in FBAR. This minimizes the dissipation in the liquid and hence improves the resonator's Q.

An interesting design of an FBAR without significant reduction of Q factor in liquid environments is illustrated in Fig. 4 (Pottigari et al., 2009). This device employs a thin vacuum gap between the FBAR and the sensing interface to prevent acoustic energy loss in liquid. This vacuum gap acts as an acoustic energy loss isolation layer of the FBAR in the aqueous viscoelastic media and reduces the direct contact area at the interface between FBAR and liquid. This vacuum separation is achieved by using micro-posts between the top electrode and the sensing diaphragm, which is in contact with the liquid. When the liquid is loaded on the sensing diaphragm, the mass is directly transferred onto the FBAR through the microposts. The vacuum gap protects the FBAR surface from the liquid, prevents the acoustic energy loss from the liquid, and contributes to maintenance of a high Q factor in the liquid without losing mass sensitivity.

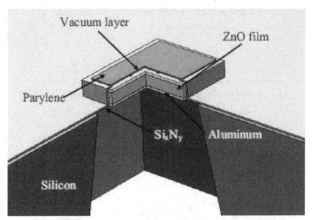

Fig. 4. Cross-section view of the V-FBAR device (Pottigari et al., 2009).
(© [2009] IEEE). Used with permission.

This vacuum (V)-FBAR uses a 0.7 µm thick ZnO film as piezoelectric material. The ZnO piezoelectric film is deposited by a radio-frequency (RF) sputtering system. A 1.6 µm thick parylene layer forms the top sensing diaphragm. Fig. 5 shows the V-FBAR integrated with a parylene sensing diaphragm that is supported by several parylene microposts. The height of the vacuum space is 2 µm and the active sensing area is 200 µm x 200 µm (Pottigari et al., 2009). The sensing diaphragm fabricated over a vacuum gap and the microposts are enhancing the sensitivity of the FBAR by separating the liquid damping effect from the operating frequency of the device. This device was conceived for biological applications.

Another recently developed MEMS based FBAR-type biological sensor is illustrated in Fig. 6 (Gabl et al., 2003). This FBAR is formed by two electrodes, fabricated at both sides of a thin a piezoelectric layer and integrated above a silicon substrate. The active vibrating region of the resonator is coated with a receptor layer which is sensitive to the biological analyte to be detected. The attachment of the analyte molecules leads to an increase of the resonator mass load and a decrease of the resonant frequency which can be electrically determined.

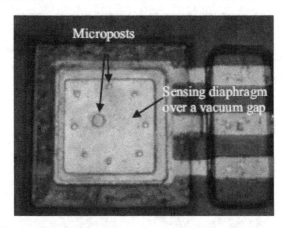

Fig. 5. Photo of fabricated V-FBAR (Pottigari et al., 2009).
(© [2009] IEEE). Used with permission.

The FBAR sensors have been fabricated on silicon substrates employing reactive magnetron sputtering of ZnO (Fig. 7). The 100 nm Au top-electrode provides a low electrical series resistance as well as a common chemical base for the binding of bioreceptor molecules. The quarter wavelength thick bottom electrode acts acoustically as an efficient reflection layer and ensures a high mass sensitivity as well as a low ohmic series resistance. A 3-fold ZnO/Pt mirror is fabricated below the FBAR sensor in order to isolate the sensor from the silicon substrate, see Figs. 6 and 7.

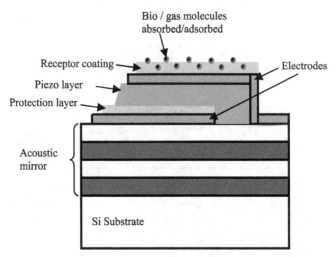

Fig. 6. Illustration of FBAR type biological sensor.

To verify the use of this type of FBAR in biosensing, a receptor assay of biotin-labeled DNA oligos coupled to the gold surface and streptavidin as the target molecule, were used. This FBAR biosensor is characterized by a sensitivity three orders of magnitude larger than for typical QCM.

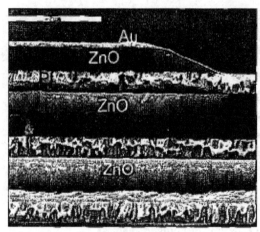

Fig. 7. SEM image of the acoustic mirror and ZnO thin films on a Si substrate. (Gabl et al., 2003). (© [2003] IEEE). Used with permission.

A nano-FBAR biosensor that uses nanomaterials to increase its sensitive area and it is based on $Mg_xZn_{1-x}O$ as piezoelectric material was reported in the literature (Chen et al., 2009). This device is built on Si substrates with an acoustic mirror consisting of alternating quarter-wavelength silicon dioxide (SiO_2) and tungsten (W) layers to isolate the FBAR from the Si substrate, see Figs. 8 and 9. High-quality ZnO and $Mg_xZn_{1-x}O$ thin films are achieved using RF sputtering technique. Tuning of the device's operating frequency was realized by varying the Mg composition in the piezoelectric $Mg_xZn_{1-x}O$ layer. The ZnO nanostructures were grown on Au electrode situated on the TFBAR's top surface using metalorganic

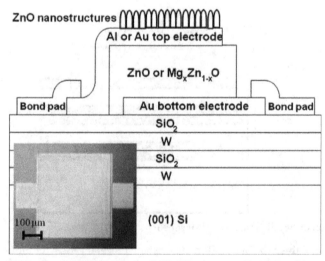

Fig. 8. Schematic diagram of the MgxZn1-xO FBAR structure; the inset shows a micrograph of a FBAR device. For the FBAR nanosensor, Au top electrode is used to facilitate MOCVD growth of ZnO nanostructures (Chen et al., 2009). (© [2009] IEEE). Used with permission.

chemical vapor deposition (MOCVD). These nanostructures offer a very large sensing area, faster response, and higher sensitivities over the planar sensor configuration. Because of the large sensitive area, the mass sensitivity of this biosensor is higher than 103 Hz cm²/ng. In order to employ this nano-FBAR for biosensing, the nanostructured ZnO surface was functionalized to selectively immobilize DNA. The device sensitivity was 16.25 ng of hybridized DNA and linker molecules combined.

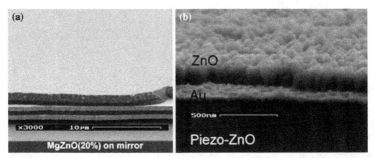

Fig. 9. Cross-sectional SEM images of (a) Mg0.2Zn0.8O film deposited on the mirror/Si structure and (b) ZnO nanostructures deposited on Au electrodes (Chen et al., 2009). (©[2009] IEEE). Used with permission.

The FBAR can be modeled into an equivalent circuit using a constant "clamped" capacitance C_0 connected in parallel with an acoustic (motional) arm that consists of motional capacitance C_m, motional inductance L_m, and motional resistance R_m, as illustrated in Fig. 10. The formulas for each component are given below:

$$C_0 = \varepsilon_r \varepsilon_0 \frac{A}{d},$$
(6)

where A is the area of overlap of the two electrodes, ε_r is the relative static permittivity of the material between the electrodes, ε_0 is the electric constant and d is the separation between the electrodes.

$$C_m = \left[\left(\frac{f_p}{f_s} \right)^2 - 1 \right]$$
(7)

where f_p is the parallel resonant frequency and f_s is the series resonant frequency.

$$L_m = \frac{1}{\left(2\pi f_s \right)^2 C_m}$$
(8)

$$R_m = \frac{1}{2\pi f_s C_m Q}$$
(9)

where Q is the quality factor of the resonator.

The FBAR biosensor is operated based on the dependency of the resonator frequency on its mass. The biomolecules attached on the sensing electrode increase the sensor mass. The sensitivity S (frequency shift per mass attachment) is described as:

$$S = \frac{f_0}{M}$$ (10)

where f_0 is the operating frequency and M is the resonator mass [11].

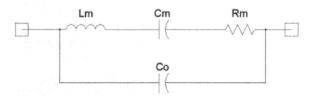

Fig. 10. FBAR equivalent circuit, where L_m is the motional mass, C_m is the motional capacitance, R_m is the motional resistance and C_0 is the static capacitance.

2.2 SAW delay lines as biosensors
Surface acoustic wave (SAW) delay lines were also studied for biosensing and could be integrated in microfluidic systems. A SAW delay line consists of two IDTs that are electrode pairs fabricated on the same side of a thin piezoelectric layer via photolithography. A sinusoidal voltage applied to the input IDTs translates into oscillating mechanical strain that forms a SAW that propagates along the surface of the piezoelectric thin film. The SAW is then

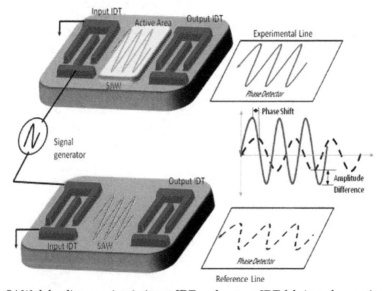

Fig. 11. A SAW delay line consists in input IDT and output IDT fabricated on a piezoelectric substrate. Two delay lines operate in parallel, with one line acting as a reference line and the other acting as an experimental line. A sinusoidal voltage is applied to the input IDT, which develops an alternating electric field that is translated into a mechanical SAW by the piezoelectric effect. The velocity of the SAW is affected by the mass loading, the liquid viscosity and the temperature of the substrate surface. Any difference in velocity between the two delay lines will be reflected as a phase shift and amplitude difference.

converted back into a sinusoidal voltage of different frequency (phase) and amplitude at the output IDTS. These differences are related to changes in the velocity of the SAW and can be correlated to changes in the mass loading, viscosity and temperature of the substrate. The effect of the temperature on the substrate could be compensated by using a dual delay line configuration. In this case only one of the SAW devices will be functionalized with biological molecules. The reference SAW delay line is not functionalized with biological molecules and it is used only for temperature compensation. Both SAW delay lines will operate at the same temperature. Figure 11 illustrates a SAW biosensor in a dual delay line configuration (Arruda et al., 2009). Measurements comparing the experimental delay line with the reference delay line are used to compensate the effect of temperature on the biosensor.

A layer of bio-molecules consisting of protein cross-linkers and antibodies is coated on the sensitive surface of the sensing device in the path of the traveling waves, between the two sets of IDTs, as illustrated in Fig. 11. If specific target proteins (antigens) are present, they bind to the antibodies, creating the mass loading on the surface of the substrate. As a result, a time delays in the propagation of the SAWs will occur.

2.3 Guided Surface Acoustic Wave resonators

One subcategory of SAW, the Love propagation mode, is especially promising for chemical and biosensing applications due to its high sensitivity to mass loading and the ability to make measurements in liquid environments with minimal propagation losses. Shear-horizontal waves can be guided by placing a thin guiding layer on a SH-SAW sensor. The bare SH-SAW resonator has a lower sensitivity because the acoustic wave goes deeper into the substrate. The sensitivity to surface perturbations could be increased when a waveguiding layer fabricated on top of the IDTs is used. The waveguiding layer also provides protection from chemicals in the liquid. Waves that propagate through the guiding layer are known as Love waves. Dielectrics such as silicon dioxide, silicon nitride and most polymers can be used as waveguide materials. Polymers have a lower shear wave velocity and therefore they are recommended for Love mode SAW sensors. Acoustic efficiency is improved when the waveguide layer is thin and does not load or attenuate the traveling acoustic wave. A typical Love mode SAW sensor is illustrated in Fig. 12.

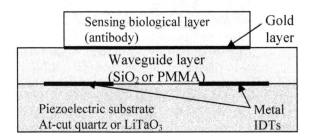

Fig. 12. Illustration of a Love mode surface acoustic wave sensor formed by a piezoelectric substrate, the waveguide layer, IDTs, and the sensing layer. The waveguide layer could be a SiO_2 film or a polymer layer and its function is to minimize propagation losses.

An example of a MEMS Love wave SAW device was implemented on $LiTaO_3$ substrates with Cr/Au IDTs and poly(methyl methacrylate) (PMMA) waveguide (Bender et al., 2000;

Branch et al., 2004). The waveguide is used to trap the SAW along the surface of the piezoelectric substrate to minimize energy losses and to protect the IDTs from corroding in a liquid-sensing environment. The surface of the waveguide is in contact with the sensing layer and it is often covered with a 50 nm thick gold layer, as illustrated in Fig.12. This gold layer provides better adherence of the sensing layer (antibodies) and prevents the nonspecific binding of proteins to the active area. Fabrication of the Cr/Au IDTs used standard lithographic techniques and the PMMA layer was spin-coated on the device. The thickness of the PMMA layer was optimized to avoid acoustic attenuation. These devices were tested with goat immunoglobulin G (IgG) and indicated minimum mass sensitivity of 17 pg/mm^2.

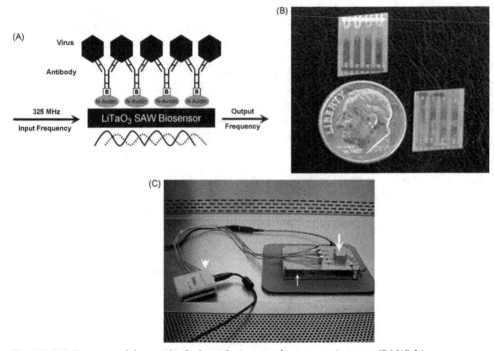

Fig. 13. (A) Concept of the antibody-based virus surface acoustic wave (SAW) biosensor. The lithium tantalate (LiTaO3) sensor surface was coated with NeutrAvidin Biotin Binding Protein (N-Avidin) and coupled to biotinylated (B) monoclonal anti-JVB antibody for Coxsackie virus. Anti-SNV-G1 glycoprotein scFv antibody for SNV detection was coupled directly. Molecular interaction between virus and antibody elicits an acoustic wave leading to a change in the input frequency of 325 MHz. (B) The sensor wafer is shown in scale compared to a dime coin; four aluminum delay lines are visible; one serves as the reference and three as the test delay lines. (C) The SAW detection board (thin arrow) with the fluidic housing (thick arrow) and the output interface device (arrowhead) to a laptop computer is shown. Reprinted with permission from Ref. (Bisoffi et al., 2008).
(© [2008] Biosensors and Bioelectronics). Used with permission.

Love wave SAW biosensors have also been used to test specific binding of different concentrations of Immunoglobulin G in the range of 0.7- 667 nM using a sensing surface modified with protein A (Gizeli et al., 2003). In this case three different Love wave SAW resonators with different piezoelectric substrate were used: (1) a LiTaO₃ substrate operating at 104 MHz, (2) a quartz substrate operating at 108 MHz and (3) a quartz substrate operating at 155 MHz. This biosensor employs polymer waveguides fabricated using simple spin coating methods (Gizeli et al., 2003). Results indicate that the thickness of the polymer guiding layer is critical factor that determines the maximum sensitivity for a given geometry. It was also found that increasing the frequency of operation results in a further increase in the device sensitivity to protein detection.

A Love wave SAW sensor with a sensing layer of anti-bacillus antibodies was used to detect low levels of bacillus thuringiensis in aqueous conditions. Tests using bovine serum albumin (BSA) in place of B. thuringiensis spores indicated a detection limit of 0.187 ng BSA (Branch et al., 2004). When applied to the detection of bacteria in food and water, SAW biosensors allow rapid, real-time, and multiple analyses, with the additional advantages of their cost effectiveness and simplicity. The drawbacks associated with this kind of biosensors include relatively long incubation times of the bacterial sample on the biosensor surface, problems with crystal surface regeneration, high packaging cost, and difficulties to implement the related microfluidic system.

A lithium tantalate-based Love wave SAW transducer with silicon dioxide waveguide sensor platform featuring three test and one reference delay lines was used to adsorb antibodies directed against either Coxsackie virus B4 or the category A bioagent Sin Nombre virus (SNV), a member of the genus Hantavirus, family *Bunyaviridae*, negative-stranded RNA viruses (Bisoffi et al., 2008). The biosensor was fabricated using metal evaporation, plasma enhanced chemical vapor deposition (PECVD), and RIE techniques on a 36° y-cut, x-propagating lithium tantalate (LiTaO₃) wafer. This biosensor, illustrated in Fig. 13, was able to detect SNV at doses lower than the load of virus typically found in a human patient suffering from hantavirus cardiopulmonary syndrome (HCPS). Further, in a proof-of-principle real world application, this Love wave SAW biosensor was capable to selectively detect SNV agents in complex solutions, such as naturally occurring bodies of water (river, sewage) without analyte preprocessing.

3. Acoustic wave MEMS devices used for telecommunications

The explosive growth of the telecommunications industry in the recent decades has created a demand for high quality, compact and mobile radio frequency (RF) modules (Reindl et al., 1996). These mobile terminals typically consist of RF integrated circuits (RFICs) and a multitude of passive components. Driven by the success of the wireless technology business and marked progress in the submicron semiconductor fabrication techniques, acoustic wave technology has progressed to GHz range in recent years (Reindl et al., 1996). Acoustic wave devices in telecommunications are typically named according their acoustic wave propagation modes; bulk and surface. Bulk acoustic wave devices have the acoustic waves propagating through the substrate. Surface acoustic wave (SAW) resonators have the acoustic waves propagating along the top surface of the device. In this section we illustrate current activities in the acoustic wave resonators for both film and surface modes.

SAW devices are technologically mature compared to bulk wave devices. The invention of thin-film interdigital transducer (IDT) as a means of manipulating surface acoustic waves (SAW) in 1965 by White and Voltmer (Visser et al., 1989; White et al., 1965) has led to the development of numerous complex electronic signal processing devices. This ingenious idea allowed a method of controlling electronic signals by transforming them into acoustic waves and manipulating the waves using patterns on piezoelectric substrates (Campbell, 1998; Morgan, 1985). Using the basic theoretical description of surface acoustic wave propagation presented by Lord Raleigh and the photolithographic techniques of microelectronics which are capable of fabricating small-size IDTs, a proliferation of radio-frequency (RF) SAW devices have been designed with innovative applications in wireless communications, radar and broadcasting systems (Hunter et al., 2002; Reindl et al., 1996; Weigel et al., 2002). Among examples of fabricated SAW devices are as components in satellite receivers, remote control units, keyless entry systems, television sets to identification tags (Campbell, 1998; Hikita et al., 2000; Springer et al., 1998). Other emerging applications of SAW resonators include gas sensors (Sadek et al., 2006), biosensors (Z. Xu, 2009), chemical (Nomura et al., 1998), temperature and pressure sensors (Buff et al., 1997). The significance of these devices in this industry can be measured in numbers by the worldwide production of these devices, where approximately 3 billion acoustic wave filters are used annually, primarily in mobile cell phones and stations (Reindl et al., 1996).

One of the most important applications of SAW resonators in telecommunication systems are as oscillators. In general, there are two major categories of MEMS resonators used in oscillator circuits namely 1) Purely mechanical resonator or flexural mode resonator can be formed using a beam or disk which will vibrate at a specific frequency and 2) Electro-mechanical resonators, usually formed using a piezoelectric membrane which is electrically excited to produce a traveling mechanical acoustic wave at a specific resonant frequency. The performance of the resonators is highly dependent both on the structure and design parameters of the resonator (Rebeiz, 2003). The essential requirements of a resonator with wireless applications include having precise resonant frequencies (f_r, f_p), low insertion losses, and high quality factors (Q) in the range of 10,000s (Zhou, 2009). Motivation of such devices is two-fold, they have powerful signal processing capabilities through control of surface waves and they are easily manufactured. Fabrication of SAW resonators typically requires a single deposition step, in comparison to their more fabrication-complex film bulk acoustic wave resonators counterparts (Ruby et al., 2001).

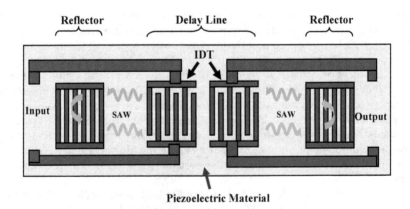

Fig. 14. Schematic of a Surface Acoustic Wave resonator

A typical SAW device (shown in Fig. 14) is composed of a piezoelectric substrate with thin-film metallic structures such as IDTs and reflectors deposited on top of the substrate's surface (Reindl et al., 1996). The operating principle of a SAW device is based on the piezoelectric effect where an applied microwave voltage input at the transmitting (input) IDT generates a propagating acoustic wave on the surface of the substrate (Campbell, 1998; Morgan, 1985; Reindl et al., 1996). This propagating acoustic wave in turn produces an electric field localized at the surface which can be detected and translated back into an electrical signal at the output IDT port (Morgan, 1985). Different from SAW delay-lines which operate based on traveling acoustic waves, SAW resonators operate using standing waves. Standing waves, or resonance, are created by the presence of reflectors, which contain the acoustic waves within the cavity. The array of metal strips or reflectors minimizes losses by reflecting and containing the acoustic waves within the cavity, thereby reducing the losses of the waves propagating outwards (Morgan, 1985).

SAW devices are typically fabricated on piezoelectric substrates and are packaged as discrete components. Very little silicon integration is achieved since these devices are connected to their CMOS counterparts on printed circuit boards, resulting in an overall large footprint of the system. To match the CMOS-based RF-circuitry and to realize a single-chip transceiver system, there have been efforts to integrate SAW devices on silicon. One such example is illustrated in (Nordin et al., 2007) where a CMOS SAW resonator was fabricated using 0.6 μm AMIs standard CMOS technology process with additional MEMS post-processing. The cross-section of the device is shown in Fig. 15. The SAW transducers are placed underneath the piezoelectric thin film to allow better CMOS compatibility. With this topology, the SAW transducers can be fabricated using the metal structures of the standard CMOS process. Thin metal wires of submicron width are common structures for fabrication of integrated circuits using the standard CMOS process. The MEMS SAW resonator can be placed beside the CMOS RF-circuits such as amplifiers to form an oscillator. Interconnections between the SAW resonator and the active circuits can be done using internal metal layers, reducing the parasitic effects of lossy, external bond wires and allows both the resonator and circuits to be placed on the same chip. This CMOS SAW resonator is still in the developmental stage and measured Q factors were less than 500.

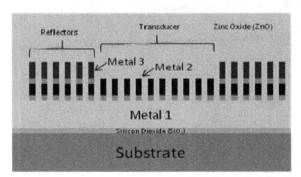

Fig. 15. Cross section of CMOS SAW resonator with increased height reflectors.

BAW resonators have acoustic waves propagating through the piezoelectric thin films. The metal transducers are placed on top and at the bottom of the piezoelectric material. Conductive metals such as Au, Mo or Pt can be used as electrodes. Acoustic waves can be generated an electrical signal is placed at the top electrode and detected the bottom electrode. The thickness of the material determines the resonant frequency of the BAW device as shown in (1)

$$f_r = \frac{v}{2d} \tag{1}$$

where f_r is the resonant frequency, v is the acoustic wave velocity and d is the thickness of the piezoelectric material. Using (1), a thick piezoelectric layer will have low resonant frequency and vice-versa.

Integration with CMOS circuits for FBARs is more challenging compared to SAW resonators. The metal electrodes and piezoelectric thin films have to be deposited separately on silicon and cannot utilize existing CMOS processes. Interconnections between the FBAR and the CMOS circuits can be made using a separate metal deposition or using bond wires. However, FBARs are more attractive compared to SAWRs due to their high Qs (10,000) and low insertion losses. As such, a lot of work has been done to integrate silicon-based FBARs with CMOS circuits (Campanella et al., 2008; Hara et al., 2003; Otis et al., 2003; Zuo et al., 2010). Aluminum nitride is typically utilized as its piezoelectric layer due to its high coupling coefficient and silicon compatibility. An example of an AlN FBAR on Si is shown in Fig. 16 (Hara et al., 2003). Three different structures were fabricated in this work namely; the acoustic diaphragm type resonator, air gap type resonator (AGR) and solidity mounted type resonator (SMR). All three structures have acoustic isolation from the lossy silicon substrate to yield higher Qs. The AGR demonstrated superior experimental results of Q = 780 and an effective electromechanical coupling constant (k_{eff}) of 5.36 % compared to the other two structures.

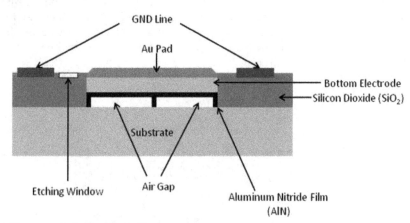

Fig. 16. Cross-section of a film bulk acoustic wave resonator

To improve the performance of the resonator without sacrificing silicon compatibility, hybrid SAW and BAW devices have been designed. An example of the hybrid device is the lateral-field excited (LFE) resonator, was successfully fabricated with Q factors in the order of 1000 (Zuo et al., 2010). This LFE resonator has Pt electrodes at the top and placed on Si substrates as shown in Fig. 17. The Si substrate is later back-etched to create the AlN LFE membrane. Elimination of the bottom FBAR electrode greatly relaxes the alignment requirements. The resonant frequency of the LFE resonator is dependent on the periodic spacing of the IDT or λ similar to SAW devices. Measurements indicate high electromechanical coupling coefficient of 1.20%, due to the membrane structure and the highly conductive Pt electrodes.

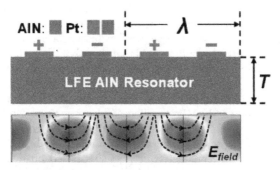

Fig. 17. Cross-section of a lateral-field-excited resonator (Zuo et al., 2010), (© [2010] IEEE). Used with permission.

Another hybrid SAW and FBAR device was reported in (Harrington et al.; Lavasani et al.) with even better Q of 6700 in air. The extensional thin-film piezoelectric-on-substrate (TPoS) shown in Fig. 18 uses molybdenum electrodes on top an AlN membrane, which is suspended above the Si substrate. The resonance frequency can be calculated using (2) as shown below:

$$2\pi f_s = (\pi/\lambda) \sqrt{(E_{eff} / \rho_{eff})} \tag{2}$$

Where E_{eff} and ρ_{eff} are the effectiveYoung's Modulus and density of the composite structure, respectively. The AlN membrane is isolated from the lossy substrate, allowing the device to have very low motional resistance of 160 Ohms. The AlN membrane and Mo electrode is anchored to the substrate on both sides. Usage of multiple anchors were also investigated and has proven to suppress the spurious modes (Harrington et al.).

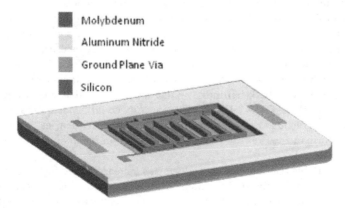

Fig. 18. Schematic for 7th order thin-film piezoelectric on substrate (TPoS) resonator.

4. Conclusion

This chapter is focused on two important applications of the acoustic-wave based MEMS devices; (1) biosensors and (2) telecommunications. Only few types of acoustic wave devices could be integrated in microfluidic systems without significant degradation of the quality factor. The acoustic wave based MEMS devices reported in the literature as biosensors are film bulk acoustic wave resonators (FBAR) and SAW resonators and delay lines. The Love mode SAW devices are often used as biosensor because the acoustic energy is confined to the sensing surface resulting in higher sensitivity to surface perturbations. The experimental results demonstrate that Love mode biosensors have high detection sensitivity.

Acoustic waves offer a promising technology platform for the development of biosensors and small-sized, low power RF-MEMS filters and resonators. MEMS acoustic wave biosensors are characterized by high sensitivity, small size and portability, fast responses, ruggedness and robustness, high accuracy, compatibility with integrated circuit (IC) technology, and excellent aging characteristics. Sensors based on this technology can be manufactured using standard photolithography and hence can be produced as relatively inexpensive devices. Integration of acoustic elements and electronic circuitry on a single silicon chip allows smart acoustic microsensors with advanced signal processing capabilities to be realized. Acoustic waves based biosensors offer the possibility of observing real-time binding events of proteins and other important biological molecules at relevant sensitivity levels and at low cost.

Acoustic wave MEMS devices used in telecommunications applications are also presented in this chapter. Telecommunication devices have different requirements compared to biosensors, where acoustic wave devices operating as a filter or resonator are expected to operate at high frequencies (GHz), have high quality factors and low insertion losses. Traditionally, SAW devices have been widely used in the telecommunications industry, however with advancement in lithographic techniques, FBARs are rapidly gaining popularity. FBARs have the advantage of meeting the stringent requirement of telecommunication industry of having Qs in the 10,000 range and silicon compatibility.

Currently, there is the concern that ZnO film, widely used as piezoelectric substrate for acoustic wave devices employed as biosensors and for telecommunications is very reactive, and unstable in liquid or air. Therefore, the stability and reliability of these devices become a problem. To solve this problem, the deposition of a thin protection layer such as Si_3N_4 on top of the ZnO film could be considered. Compared to ZnO, AlN shows a slightly lower piezoelectric coupling. However, AlN films have excellent piezoelectric properties. The Rayleigh wave phase velocity in (0 0 1) AlN is much higher than ZnO, which suggests that AlN is preferred for high frequency and high sensitivity applications (Gorla et al., 1999). AlN is a hard material with bulk hardness similar to quartz, and is chemically stable at temperatures less than about 700 °C. Therefore, using AlN could be an alternative and lead to the development of acoustic devices operating at higher frequencies, with improved sensitivity and performance in insertion loss and resistance in harsh environments (Mason et al., 1972)

The popularity of portable communication gadgets has increased the demand and necessity of small-sized, low power RF-MEMS filters and resonators. Passive acoustic wave resonators fulfill this market niche of low-power, radio frequency and silicon-compatible resonators and filters. Surface acoustic wave devices (filters and resonators) have long been popular in the communications industry. To improve silicon compatibility, efforts have been made to implement the SAW resonator using standard CMOS process with minimal post-processing. Results indicate that while this device shows promise, significant improvement is required before the CMOS SAW resonator can meet the stringent communication requirements. In this aspect, FBARs have shown better performance in terms of quality factor (6000) and low insertion losses. However, complete CMOS-compatibility has not yet been achieved and the device still requires bond wires for connections to the circuitry.

5. References

Andle, J. C., & Vetelino, J. F. (1995).*Acoustic wave biosensors*. Proceedings of IEEE Ultrasonics Symposium, ISBN 1051-0117, 7-10 Nov 1995

Arruda, D. L., Wilson, W. C., Nguyen, C., Yao, Q. W., Caiazzo, R. J., Talpasanu, I., et al. (2009). Microelectrical sensors as emerging platforms for protein biomarker detection in point-of-care diagnostics. *Expert Review of Molecular Diagnostics*, Vol. *9*, No. 7, (2009), pp. 749-755.

Ballantine, D. S., White, R. M., Frye, G. C., Martin, S. J., Ricco, A. J., Zellers, E. T., et al. (1996). *Acoustic wave sensors: theory, design, and physico-chemical applications*: Academic Press San Diego.

Bender, F., Cernosek, R. W., & Josse, F. (2000). Love-wave biosensors using cross-linked polymer waveguides on LiTaO 3 substrates. *Electronics Letters*, Vol. *36*, No. 19, (2000), pp. 1672-1673.

Bisoffi, M., Hjelle, B., Brown, D. C., Branch, D. W., Edwards, T. L., Brozik, S. M., et al. (2008). Detection of viral bioagents using a shear horizontal surface acoustic wave biosensor. *Biosensors and Bioelectronics*, Vol. *23*, No. 9, (2008), pp. 1397-1403,.

Bowers, W. D., Chuan, R. L., & Duong, T. M. (1991). A 200 MHz surface acoustic wave resonator mass microbalance. *Review of scientific instruments*, Vol. *62*, No. (1991), pp. 1624.

Branch, D. W., & Brozik, S. M. (2004). Low-level detection of a Bacillus anthracis simulant using Love-wave biosensors on 36°YX LiTaO3. *Biosensors and Bioelectronics*, Vol. *19*, No. 8, (2004), pp. 849-859.

Buff, W., Rusko, M., Goroll, E., Ehrenpfordt, J., & Vandahl, T. (1997).*Universal pressure and temperature SAW sensor for wireless applications*. Proceedings of IEEE Ultrasonics Symposium.

Campanella, H., Cabruja, E., Montserrat, J., Uranga, A., Barniol, N., & Esteve, J. (2008). Thin-Film Bulk Acoustic Wave Resonator Floating Above CMOS Substrate. *Electron Device Letters, IEEE*, Vol. *29*, No. 1, (2008), pp. 28-30.

Campbell, C. (1998). *Surface acoustic wave devices for mobile and wireless communications*. San Diego: Academic Press.

Cavic, B. A., Hayward, G. L., & Thompson, M. (1999). Acoustic waves and the study of biochemical macromolecules and cells at the sensor-liquid interface. *Analyst*, Vol. *124*, No. 10, (1999), pp. 1405-1420, ISSN 0003-2654.

Cheeke, J. D. N., Tashtoush, N., & Eddy, N. (1996).*Surface acoustic wave humidity sensor based on the changes in the viscoelastic properties of a polymer film*. Proceedings of IEEE Ultrasonics Symposium, ISBN 1051-0117, 3-6 Nov 1996.

Chen, Y., Reyes, P. I., Duan, Z., Saraf, G., Wittstruck, R., Lu, Y., et al. (2009). Multifunctional ZnO-Based Thin-Film Bulk Acoustic Resonator for Biosensors. *Journal of Electronic Materials*, Vol. *38*, No. 8, (2009), pp. 1605-1611.

Cullen, D. E., & Montress, G. K. (1980).*Progress in the Development of SAW Resonator Pressure Transducers*. Proceedings of IEEE Ultrasonics Symposium, 1980.

Cullen, D. E., & Reeder, T. M. (1975).*Measurement of SAW Velocity Versus Strain for YX and ST Quartz*. Proceedings of IEEE Ultrasonics Symposium, 1975.

Gabl, R., Green, E., Schreiter, M., Feucht, H. D., Zeininger, H., Primig, R., et al. (2003).*Novel integrated FBAR sensors: a universal technology platform for bio- and gas-detection*. Proceedings of IEEE Sensors, 22-24 Oct. 2003.

Gizeli, E., Bender, F., Rasmusson, A., Saha, K., Josse, F., & Cernosek, R. (2003). Sensitivity of the acoustic waveguide biosensor to protein binding as a function of the waveguide properties. *Biosensors and Bioelectronics*, Vol. *18*, No. 11, (2003), pp. 1399-1406.

Gorla, C. R., Emanetoglu, N. W., Liang, S.,. Mayo, W. E., Lua Y., Wraback, M., & Shen, H. (1999) Structural, optical, and surface acoustic wave properties of epitaxial ZnO films grown on [011-2] Sapphire by metalorganic chemical vapor deposition", *J. Appl. Phys.* 85(51), pp. 2595, 1999.

Hara, M., Kuypers, J., Abe, T., & Esashi, M. (2003).*MEMS based thin film 2 GHz resonator for CMOS integration*. Proceedings of Microwave Symposium Digest, 2003 IEEE MTT-S International, ISBN 0149-645X.

Harrington, B. P., Shahmohammadi, M., & Abdolvand, R.*Toward ultimate performance in GHZ MEMS resonators: Low impedance and high Q*. Proceedings of IEEE 23rd International Conference on Micro Electro Mechanical Systems (MEMS), ISBN 1084-6999, 24-28 Jan. 2010

Hikita, M., Takubo, C., & Asai, K. (2000). New high performance SAW convolvers used in high bit rate and wideband spread spectrum CDMA communications system. *IEEE Transactions on Ultrasonics, Ferroelectrics and Frequency Control*, Vol. *47*, No. 1, (2000), pp. 233-241.

Hunter, I. C., Billonet, L., Jarry, B., & Guillon, P. (2002). Microwave filters-applications and technology. *IEEE Transactions on Microwave Theory and Techniques*, Vol. *50*, No. 3, (2002), pp. 794-805.

Janshoff, A., Galla, H. J., & Steinem, C. (2000). Piezoelectric Mass-Sensing Devices as Biosensors-An Alternative to Optical Biosensors? *Angew Chem Int Ed Engl*, Vol. *39*, No. 22, (2000), pp. 4004-4032, ISSN 1521-3773.

Länge, K., Rapp, B. E., & Rapp, M. (2008). Surface acoustic wave biosensors: a review. *Analytical and Bioanalytical Chemistry*, Vol. *391*, No. 5, (2008), pp. 1509-1519.

Lavasani, H. M., Wanling, P., Harrington, B., Abdolvand, R., & Ayazi, F. A 76 dB Ohm 1.7 GHz 0.18 um CMOS Tunable TIA Using Broadband Current Pre-Amplifier for High Frequency Lateral MEMS Oscillators. *Solid-State Circuits, IEEE Journal of*, Vol. *46*, No. 1, pp. 224-235.

Levit, N., Pestov, D., & Tepper, G. (2002). High surface area polymer coatings for SAW-based chemical sensor applications. *Sensors and Actuators B: Chemical*, Vol. *82*, No. 2-3, (2002), pp. 241-249.

Macchiarella, G., & Stracca, G. B. (1982).*SAW Devices for Telecommunications: Examples and Applications*. Proceedings of IEEE Ultrasonics Symposium, 1982

Mason, W. P., Thurston, R. N., (1972) *Physical Acoustics*, Academic Press, Inc., New York, 1972.

Morgan, D. P. (1985). *Surface-wave devices for signal processing*. Amsterdam ; New York: Elsevier.

Nakamoto, T., Nakamura, K., & Moriizumi, T. (1996).*Study of oscillator-circuit behavior for QCM gas sensor*. Proceedings of IEEE Ultrasonics Symposium, ISSN 1051-0117, 3-6 Nov 1996

Nomura, T., Takebayashi, R., & Saitoh, A. (1998). Chemical sensor based on surface acoustic wave resonator using Langmuir-Blodgett film. *IEEE Transactions on Ultrasonics, Ferroelectrics and Frequency Control*, Vol. 45, No. 5, (1998), pp. 1261-1265.

Nordin, A. N., & Zaghloul, M. E. (2007). Modeling and Fabrication of CMOS Surface Acoustic Wave Resonators. *IEEE Transactions on Microwave Theory and Techniques*, Vol. 55, No. 5, (2007), pp. 992-1001.

Otis, B. P., & Rabaey, J. M. (2003). A 300-/spl mu/W 1.9-GHz CMOS oscillator utilizing micromachined resonators. *IEEE Journal of Solid-State Circuits*, Vol. 38, No. 7, (2003), pp. 1271-1274.

Pohl, A., Ostermayer, G., Reindl, L., & Seifert, F. (1997).*Monitoring the tire pressure at cars using passive SAW sensors*. Proceedings of IEEE Ultrasonics Symposium, 1051-0117, 5-8 Oct 1997.

Pottigari, S. S., & Jae Wan, K. (2009).*Vacuum-gapped film bulk acoustic resonator for low-loss mass sensing in liquid*. Proceedings of International Solid-State Sensors, Actuators and Microsystems Conference, 21-25 June 2009

Rebeiz, G. M. (2003). *RF MEMS: theory, design, and technology*: John Wiley and Sons.

Reindl, L., Scholl, G., Ostertag, T., Ruppel, C. C. W., Bulst, W. E., & Seifert, F. (1996).*SAW devices as wireless passive sensors*. Proceedings of IEEE Ultrasonics Symposium, 1051-0117, 3-6 Nov 1996

Ruby, R. C., Bradley, P., Oshmyansky, Y., Chien, A., & Larson, J. D., III. (2001).*Thin film bulk wave acoustic resonators (FBAR) for wireless applications*. Proceedings of IEEE Ultrasonics Symposium,

Sadek, A. Z., Wlodarski, W., Shin, K., Kaner, R. B., & Kalantar-zadeh, K. (2006). A layered surface acoustic wave gas sensor based on a polyaniline/In2O3 nanofibre composite. *Nanotechnology*, Vol. 17, No. 17, (2006), pp. 4488, ISSN 0957-4484

Skládal, P. (2003). Piezoelectric quartz crystal sensors applied for bioanalytical assays and characterization of affinity interactions. *Journal of the Brazilian Chemical Society*, Vol. 14, No. (2003), pp. 491-502, ISSN 0103-5053

Smith, A. L. (2001). Mass and heat flow measurement sensor: Google Patents.

Springer, A., Huemer, M., Reindl, L., Ruppel, C. C. W., Pohl, A., Seifert, F., et al. (1998). A robust ultra-broad-band wireless communication system using SAW chirped delay lines. *IEEE Transactions on Microwave Theory and Techniques*, Vol. 46, No. 12, (1998), pp. 2213-2219.

Staples, E. J. (1999).*Electronic nose simulation of olfactory response containing 500 orthogonal sensors in 10 seconds*. Proceedings of IEEE Ultrasonics Symposium, 1051-0117, 1999.

Vellekoop, M. J., Nieuwkoop, E., Haartsan, J. C., & Venema, A. (1987). *A Monolithic SAW Physical-Electronic System for Sensors*. Proceedings of IEEE Ultrasonics Symposium, 1987

Vellekoop, N. J., Jakoby, B., & Bastemeijer, J. (1999).*A Love-wave ice detector*. Proceedings of IEEE Ultrasonics Symposium, ISBN 1051-0117, 1999

Vetelino, K. A., Story, P. R., Mileham, R. D., & Galipeau, D. W. (1996). Improved dew point measurements based on a SAW sensor. *Sensors and Actuators B: Chemical*, Vol. 35, No. 1-3, (1996), pp. 91-98,

Visser, J. H., Vellekoop, M. J., Venema, A., Drift, E. v. d., Rek, P. J. M., & Nederhof, A. J. (1989).*Surface Acoustic Wave filter in ZnO-SiO2-Si layered structures*. Proceedings of IEEE Ultrasonics Symposium.

Weber, J., Albers, W. M., Tuppurainen, J., Link, M., Gabl, R., Wersing, W., et al. (2006). Shear mode FBARs as highly sensitive liquid biosensors. *Sensors and Actuators A: Physical*, Vol. 128, No. 1, (2006), pp. 84-88.

Weigel, R., Morgan, D. P., Owens, J. M., Ballato, A., Lakin, K. M., Hashimoto, K., et al. (2002). Microwave acoustic materials, devices, and applications. *IEEE Transactions on Microwave Theory and Techniques*, Vol. 50, No. 3, (2002), pp. 738-749.

Weld, C. E., Sternhagen, J. D., Mileham, R. D., Mitzner, K. D., & Galipeau, D. W. (1999).*Temperature measurement using surface skimming bulk waves*. Proceedings of IEEE Ultrasonics Symposium, 1051-0117, 1999.

White, R. M., & Voltmer, F. W. (1965). Direct piezoelectric coupling to surface elastic waves. *Applied Physics Letters*, Vol. 7, No. (1965), pp. 314-316,

Wingqvist, G., Bjurstrom, J., Liljeholm, L., Katardjiev, I., & Spetz, A. L. (2005).*Shear mode AlN thin film electroacoustic resonator for biosensor applications*. Proceedings of IEEE Sensors, Oct. 30 2005-Nov. 3 2005

Wohltjen, H., & Dessy, R. (1979). Surface acoustic wave probe for chemical analysis. I. Introduction and instrument description. *Analytical Chemistry*, Vol. 51, No. 9, (1979), pp. 1458-1464.

Zhang, X., Xu, W., Abbaspour-Tamijani, A., & Chae, J. (2009).*Thermal Analysis and Characterization of a High Q Film Bulk Acoustic Resonator (FBAR) as Biosensors in Liquids*. Proceedings of IEEE 22nd International Conference on MEMS, pp. 939-942.

Zhou, W. (2009). *Integration of MEMS Resonators within CMOS Technology*. Cornell.

Zuo, C., Van der Spiegel, J., & Piazza, G. (2010). 1.05-GHz CMOS Oscillator Based on Lateral-Field-Excited Piezoelectric AlN Contour-Mode MEMS Resonators. *Chengjie Zuo*, Vol., No. (2010), pp. 15.

4

MEMS Microfluidics for Lab-on-a-Chip Applications

Nazmul Islam and Saief Sayed
MEMS/NEMS Lab, The University of Texas at Brownsville
USA

1. Introduction

Micro-/Nano- fluid devices are becoming more prevalent, both in commercial applications and in scientific inquiry. Microfluidics, a branch of MEMS (Micro-Electro-Mechanical Systems) is key enabling factor in the miniaturization and integration of multiple functionalities for chemical analysis and synthesis in handheld microdevices, which require efficient methods for manipulating ultra small volumes of liquid as well as the contents in the fluid within the fluid networks. For biomedical applications, microfluidic chip arrays are being used to identify multiple bioparticles [1]. Recent developments in micro-fabrication technologies enabled different types of microfluidic functions such as micro-pumps [2, 3], micro-mixers [4], particle concentrator [5, 6], and various s types of injection systems (nano-needles). At the very beginning of microfluidics, people thought that microfluidic devices could just be a miniaturized version of macro- fluidic devices. The technological advancement on microfluidic systems has proven that the problem is far more complicated than scaling down a device geometrically. Therefore, a better understanding of the micro/nano scale properties is in order.

A dominant difference of microfluidic devices from their macro-scale counterparts is the increased surface/volume ratio, hence dominant surface force effects/friction. Micro channel needs high pressure for pressure driven flow to produce sufficient flow rate. The formula below relates the applied pressure with the conduit radius for a constant flow rate.

$$\Delta P = \frac{8\mu L Q}{\pi a^4} \qquad \mu : \text{viscosity;} \quad a : \text{conduit radius}$$

Every time we try to reduce the conduit radius into half, we need to have sixteen times of larger pressure to sustain the same flow rate. So at microscale, surface forces start to dominate due to the large surface/volume ratio. Therefore electroosmosis (as a type of surface forces) becomes the prime candidates for fluidic manipulation at micro scale. Direct Current Electroosmosis (DCEO) has a long history of being applied in miniaturized biochemical devices. However, DCEO has many undesirable effects, such as high voltage operation, electrolysis and resulting bubble generation, and pH gradient. In this chapter, we examined a new type of EO phenomena, ACEO (Alternating Current Electroosmosis), and how it can be employed to integrate with the microcantilever particle trapping.

2. MEMS microfluidics: Past, present and future

Microfluidics is an interdisciplinary area that focuses on the miniaturization of fluid-handling systems. The concept of complete lab-on-a-chip devices or micro-total analysis systems (μ-TAS), has recently generated great interest in a variety of industries, where transport and processes (including mixing, reaction, separation, and manipulation of chemicals and particles) are being applied on much smaller scales than traditional engineering technologies [7, 8]. This interest has led to tremendous growth in microfluidic technologies over the past decade.

A functional microfluidic chip should be able to realize certain functions, such as transporting, mixing (with reactants), sample treatment (concentrating, sorting). Figure 1 gives an example of a generic microfluidic chip. Unlike the microelectronics industry, where the current emphasis is on reducing the size of transistors, the field of microfluidics is focusing on investigating new fluid phenomenon at micro/nano-scale with more sophisticated fluid-handling capabilities. One of the promising types of micropumps is driven by electroosmosis (EO). EO pumps are purely driven by electric fields and have no moving parts. The central concept is utilizing the surface force. As the surface to volume ratio of the microchannel is high, the dominant surface force is a good choice for pumping the liquid in microchannel.

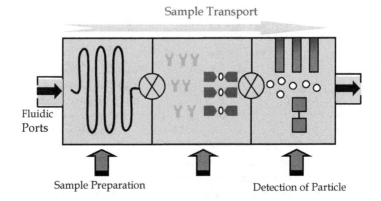

Fig. 1. A Generic Microfluidic Chip with the sample preparation, selection and detection of bio/nano/micro-particles.

As mentioned before, the field of microfluidics has far more complexity than people first expected. The dominant/efficient mechanism to manipulate fluid or biochemical samples will change with sample conductivity, pH value, and sample sizes. Also, we need to be concerned with side effects from those actuation mechanisms. So microfluidic industry did not develop in a similar way as the microelectronics industry. Lab-on-a-chip (LOC) devices have shown commercial success in biological applications such as electrophoretic separations and DNA sequencing, where DC electrokinetics or DC electric field is used to manipulate fluid/particles (here surface force is used versus pressure driven). However, because of its operation with high voltage, there are obstacles to extend DC electrokinetics to more fluidic functions. On the other hand, AC electrokinetics has important potentials in the field of life science. With the capability to manipulate particle and fluid motion at the microscale with low voltage, it meshes well with the requirements of lab-on-a-chip systems.

In this chapter a combination of sample transportation and particle detection in a microfluidic chip. As shown in figure 1, we need to transport the sample, for which I have experimentally validate the biased ACEO micro-pumping. The developed micropump is operated with smaller AC voltage, which is compatible with the lab-on-a-chip. For particle detection as shown in figure 1, we need high effective techniques to detect micro/nano-scale particulate. We also have envisioned to develop the validation technique for particle trapping. That is the reason we have interface the microcantilever with our microfluidic device. The applications of this integration can greatly benefit the advancement of AC electrokinetics.

3. Electrokinetics

Electrokinetics is the combination of "electric" plus "kinetics". Generally speaking, electrokinetics is the motion of liquid or particle under the influence of electric field. According to Probstein [7], the electrokinetic effects have been first observed by F. F. Reuss in 1809 via experiments on porous clay diaphragms. He has shown that in capillary, fluid moves from anode to cathode in the presence of an external electric field. In the mid 19th century, Wiedemann repeated this experiment and described the fundamentals of electrokinetics. This was followed later by the seminal work of Helmholtz in 1879 on the electric double layer theory, which related the electrical and flow parameters for electrokinetic transport. DC electrokinetics, including electroosmosis and electrophoresis, has almost two hundreds years' history and has been rather thoroughly investigated. Electroosmosis is the fluid motion caused by the electrical force acting on the double layers next to a charge surface. The ions in a double layer are moved by a tangential electric field, giving rise to movement of the whole double layer along the surface, which in turn puts the bulk fluid into motion through the viscous interaction.

On the other hand, AC electrokinetics has been studied for just a few years. AC electrokinetics can be classified into three categories, dielectrophoresis (DEP), the electrothermal effect (ET) and ac electroosmosis (ACEO). DEP is the force acting on the particles due to the difference in polarizability between the particles and the fluid. The electrothermal effect refers to the fluid motion caused by the interaction of electric fields and gradients of conductivity and permittivity of the fluid through Joule heating. AC electroosmosis is the fluid motion induced by moving double layers. We will focus on the AC electroosmosis, while we will discuss the effects from other electrokinetic phenomena. Basically, AC electroosmosis (ACEO) works by the same principle as DC EO. However, in ACEO, the surface charges are induced by externally applied voltages, rather than naturally occurring charges in DCEO, and consequently hundreds of time stronger. As a result, ACEO can generate flow velocity of several hundreds microns per second with a couple of volts.

AC techniques are more favorable over DC ones for following reasons: (1) low operating voltage makes it superior in terms of device portability; (2) avoids electrolysis and the resulting bubble generation; (3) minimizes pH gradient; (4) miniaturization and integration with other devices on lab-chips. We have extended the scope of ACEO by including electrochemical reactions (i.e. Faradaic charging), and then developed a new ACEO technique—asymmetric polarization ACEO. Asymmetric polarization of electrodes is achieved by combining the DC bias into AC signals over electrode pairs. Biased ACEO breaks the reflection symmetry to produce net flow in a symmetric pair of electrode. This technique adds more flexibility to the faster manipulation of bio-particulate.

3.1 AC Electrokinetics

AC electrokinetics provides a means to effectively control and manipulate particles and fluids at micro-scale. Switching electric field of AC elctrokinetics can suppress the electrolysis and hence the change of pH value at electrodes, which is inevitable in dc electrokinetics. Different from DC EO which relies on naturally induced charges, ACEO induces surface charges by applying voltage, which can be hundreds of times higher lower voltage is required to generate sufficient flow velocity. Additionally small spacing of electrodes makes it possible to reach high electric field (E) with low applied voltage (V). AC electrokinetics can operate low voltages, which is suitable for lab-on-a-chip operation.

AC electrokinetics comprises of dielectrophoresis, the thermal effect and ac electroosmosis. Dielectrophoresis (DEP) is the response of particles to the applied electric fields, and the electrothermal effect and ac electroosmosis are fluid motion caused by AC fields. Ramos et al have provided a rather comprehensive review [9] on various forces acting on micro-size particles on microelectrode arrays when electrodes are energized with ac voltages over a wide range of frequency. In subsequent work, Ramos et al [10] presented a RC model describing the frequency dependence of ACEO flow velocity by capacitive charging of the electrodes. Recently, Bazant and Squires [25, 28] also presented that AC electrokinetic phenomena can also occur for conducting particles, which they termed as Induced charged Electroosmosis (ICEO). They also predict that the velocity will scale as E^2, where E is the applied electric field. Vortices will also occur around a spherical metal because of their geometry. Bhatt [5] also reported that electrohydrodynamic effect arising from the application of alternating electric fields to patterned electrode surfaces.

3.2 AC Electroosmosis

ACEO was first investigated using a pair of planar electrodes as shown in figure 2. In figure 2, a pair of electrodes is placed parallel in an electrolyte. The first half cycle of the applied signal is shown in figure 2(a). The double layer is produced on the electrode surface like dc electroosmosis. Therefore there is a nonzero tangential component of electric field acting on the double layer. The interaction of this tangential electric field with the surface charge creates the force according to the Coulomb's law. This force is along the electrode, which in turn puts the fluid in motion. In the half cycle of the applied AC signal, the sign of the surface potential becomes reversed, as shown in figure 2(b). The ions move accordingly, keeping the sign of the double layer opposite to the potential. At the same time, the electric field also reversed its direction. So in this case both surface charge polarity and the electric field direction changed. Subsequently, the force acting on the double layer is still the same direction as in the first half cycle, keeping the fluid motion unchanged.

AC electroosmosis is much more complicated than DC electroosmosis due to two reasons: 1) The applied signal is oscillating; 2) The surface charge and the tangential field are coupled. An excellent experiment and theoretical study of AC electroosmosis can be found in [9, 10, 38]. The governing equations for the fluid motion inside the double layer is represented by following equation coupled with the continuity equation.

$$\rho_m \frac{D\vec{u}}{Dt} = -\nabla p + \mu \nabla^2 \vec{u} + \rho \vec{E} \tag{1}$$

Neglecting the term on the left hand side of equation (1) leads to,

$$-\nabla p + \mu \nabla^2 \vec{u} + \rho \vec{E} = 0 \tag{2}$$

The fluid velocity can be obtained after substituting the charge density (ρ) and the electric field distribution into equation (2). The fluid velocity at the outer boundary of the double layer is given by,

$$\vec{u}_f = -\frac{\varepsilon}{4\mu} \Lambda \nabla_s \left(|\Delta V|^2 \right) \tag{3}$$

where Λ is a factor of the double layer structure, Δ_s represents a gradient across the surface, and ΔV is the potential drop across the double layer. According to the equation (3), fluid tends to move from higher electric field region to the lower electric field region. The fluid velocity at the electrode surface is,

$$u_s = \frac{1}{8} \frac{\varepsilon V_0 \Omega^2}{\mu s \left(1 + \Omega^2 \right)^2} \tag{4}$$

where V_0 is the voltage amplitude and s is the distance from the electrode center. Ω is the nondimensional frequency,

$$\Omega = \omega \frac{\varepsilon}{\sigma} \frac{\pi}{2} \frac{s}{\lambda} \tag{5}$$

where ω is the angular frequency and λ is the debye length.

Equation (4) gives a bell-shaped velocity profile with respect to the frequency. At high and low frequencies the velocity approaches to zero. We experimentally verified this bell-shaped velocity profile in our research. Our research emphasizes on the charging processes of electrodes. By studying surface EO flows with respect to AC potential, we identified ACEO induced by electrochemical reactions (i.e. Faradaic charging), and we have developed a new ACEO technique — asymmetric polarization ACEO. Here we also explain the competition between the Faradaic and Capacitive charging process.

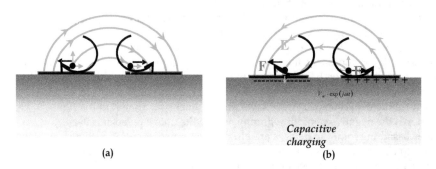

(a)　　　　　　　　　　Capacitive
　　　　　　　　　　　charging
　　　　　　　　　　　(b)

Fig. 2. ACEO fluid motion and induced charge at electrode surface. (a) during the half cycle when the left electrode has positive polarity; (b) during the next half cycle with opposite electrical polarity.

Faradaic charging generates co-ions from electrochemical reactions. When the electrode is positively charged, it goes through the reaction according to the Faradaic's law. On the other hand, capacitive charging attracts counter-ions from the electrolyte to screen the electrode potential. Our contribution in this field is that we have developed an "asymmetric polarization (A-P) ACEO" technique by adding a DC offset to the AC signal, and we used this A-P ACEO for trapping particles on the electrodes [11]. Asymmetric polarization of electrodes in a pair is achieved by combining the DC bias into AC signals over electrode pairs. For adding DC bias the reflection symmetry of electrode charging is broken, leading to asymmetric surface flow and net-flow. By adjusting the amplitude and frequency of AC signals, a variety of directed surface flows are produced on electrodes to manipulate and transport particles.

Capacitive charging and Faradaic charging coexist and compete for dominance when the electrodes are energized under biased conditions. The biased ACEO scheme is built on the fact that the two electrodes in a pair undergo the two distinct polarization processes. Biased AC EO is implemented by energizing electrode pairs with biased AC signals so that electrodes in a pair undergo polarizations different from each other. The advantage of such a scheme is twofold. First, it breaks the mirror symmetry of electric fields and, consequently, that of surface flows. If Faradaic charging occurs on one electrode while capacitive charging takes place on the other, then ions of the same sign populate the electrode surface and they migrate in the same direction under the influence of the electric fields. As a result, a unidirectional flow is produced at the electrode surface. So, non-uniform electric field has a good application in pumping action.

3.3 Fabrication of microfluidic devices

The processes developed for microelectronics, such as standard photolithographic methods, can be applied to silicon and glass substrates producing channel networks in two dimensions for sample transport, mixing, separation, and detection systems on a monolithic chip (Fig 1). A mask is made that has transparent and opaque regions that are patterned as a negative image of the desired channel layout. A UV-light source transfers the layout from the mask to the photoresist, which has been previously deposited on the substrate by spin-coating. The photoresist is then developed in a solvent that selectively removes either the exposed or the unexposed regions.

After developing the photoresist, it has a small amount of "hydrocarbon" material still on the wafer. If this is not removed it can affect the geomarty. Removing this very thin layer is something we call "descum". To descum we used an oxygen plasma generated in a parallel-plate etcher (Reactive Ion Etching – RIE). The oxygen plasma interacts with our undesirable hydrocarbon layer and burns it away. Then we deposit gold using the E-beam evaporator. As an adhesive layer we have used Cromium (Cr) between the gold and the wafer. After that photoresist is removed by lift-off process, and interdigitated electrode pattern is created for microfluidic experiment. Typical fabrication sequence is shown in the figure 3. Here we have shown the negative photolithography, where the image reversal is used.

Polymer based microfluidic devices are also getting much attention recent days. Polymer is resistant to chemicals and some are biocompatible for implantation by FDA [37]. PDMS has shown a number of advantages over other polymer materials (eg. SU-8, PMMA): PDMS microchips can be easily replicated and produced by rapid prototype approaches with low cost [31]. The excellent optical transparency of PDMS has been exploited to integrate different elements in the optical detection.

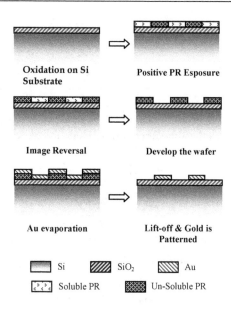

Oxidation on Si Substrate

Positive PR Esposure

Image Reversal

Develop the wafer

Au evaporation

Lift-off & Gold is Patterned

	Si		SiO₂		Au

Soluble PR   Un-Soluble PR

Fig. 3. Fabrication sequence for silicon, and glass microfluidic devices

For our experiments we have used PDMS to fabricate our microchannel for microfluidics particle trapping and micropumping. The first step of fabricating microchannel is to have the mold. We have fabricated the channel dimensions as low as 100 μm X 100 μm [cross-section]. Following is the steps of fabricating microchannel for our experiments.

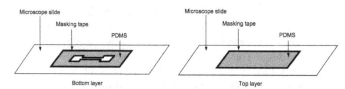

a. PDMS on two substrates to create a bottom layer and a top layer of the microchannel pump

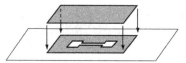

b. Top layer PDMS was peeled from the substrate after thermally cured

c. Top layer PDMS was placed on top of the patterned bottom layer

d. Top view of the sealed microchannel pump

e. Two rectangular cuts were made for reservoirs

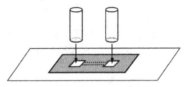

f. Integrating the reservoirs onto the pump

Fig. 4. Steps of fabricating microfluidic chamber

1. PDMS was prepared by thoroughly mixing the silicone Base and Curing Agent (SYLGARD 184 Silicone Elastomer Kit, Dow Corning, Midland, MI) at ratio 10:1 base:curing agent using a portable mixer. The mixture was then set in a small plastic container to reduce the air bubbles from the mixing process. This was done in approximately 30 minutes or until there were no visible bubbles. The PDMS was then poured onto two different glass substrates (microscope glass slides).
2. The first substrate functioned as the bottom layer, which was patterned using a piece of masking tape. The pattern defined the profile of the microfluidic pump. It contained a microchannel that connected two reservoirs at the end of the channel. The second substrate created the top layer, which would be used as the channel sealer. (See Figure 4a). Both substrates were thermally cured on a hot plate for 10 minutes at 150°C.
3. The PDMS for the top layer was then peeled from the glass substrate and was placed on top of the PDMS for the bottom layer. A careful placement was done to avoid any air gap between the two layers. Hence, to prevent a leak of the working liquid. (Figure 4-b, c, and d).
4. Two rectangular cuts were made on the top layer outlining the reservoir contours of the bottom layer. (Figure 4e).
5. Two reservoirs were made by cutting a glass pipette into two two-centimeter tubes. They were integrated on the pump using a 5-minute epoxy glue. (Figure 4f).

One of the most important advantages of PDMS is that it is easy to fabricate and allows a simple sealing with the planar substrates. However, PDMS has three significant problems in practical use.

- Dissolution or swelling by organic solvents.
- Absorption of chemical materials.
- Adhesion between PDMS and metal layer is not intact/perfect

To overcome these problems, we have coated PDMS with perfluoro amorphous polymer on the PDMS micro structures, which minimizes the problems of PDMS [36]. Perfluoro amorphous polymer (CYTOP, CTX-809A, Asahi Glass Company, Japan) is coated by spin coating on the PDMS structure. Since natural PDMS surface repels CYTOP, so we used O2 plasma pre-treatment on the surface before CYTOP coating. Our preliminary result shows that CYTOP coating was achieved without any deformation of the micro structure. This polymer based substrate with CYTOP can be used in different biological, chemical and lab-on-a-chip (LOC) applications.

The need for innovative fabrication methods to integrate higher levels of functionality into microfluidic and lab-on-a-chip devices is growing almost as rapidly as the number of potential applications for these miniature devices. The ability to make fully-integrated, multi-level fluidic systems with functional valves, pumping systems, electrical and electronic components, and other microeletromechanical system (MEMS) components is essential in order for this relatively new field to reach its full potential.

3.4 Electric field analysis

We have used Comsol Multiphysics (formerly FEMLab) to simulate the electric field distribution above a pair of planar electrodes (160micron width and 40 micron separation between the electrodes, with infinitesimal thickness). As shown in Figure 5a, tangential electric fields change directions over one electrode, which indicates that two counter-rotating vortices exist on one electrode, as schematically drawn in Figure 5b, countering to one vortex reported in the litereature.

The fluid velocity on the electrode surface is given [5] as

$$u = -(\varepsilon / \eta)(\xi - \varphi_b)E_t \tag{6}$$

where ε is the permittivity, η is the viscosity of bulk solution, $(\xi - \varphi_b)$ is the difference of potential between the double layer and the bulk solution, Et is the tangential component of electrical field. According to Equation 6, the velocity on the surface of double layer is proportional to the tangential field and potential difference between the double layer and fluid, which corresponds to the normal component of electrical field. Therefore, normalized boundary conditions on both electrodes are given as, $u = -E_x E_y$.

To model the fluid motions, the 2D Incompressible Navier-Stokes module is used in FEMLAB. In this case, the hydrodynamic property in the chamber is given as density = 1000 kg.m^{-1} and viscosity = 10^{-3} kg.m^{-1}.sec. The fluid velocity distribution is then obtained by solving Navier-Stokes equation with calculated field profile.

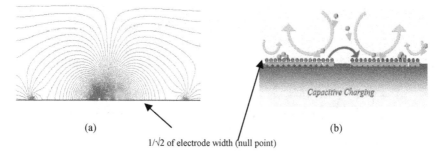

(a)   (b)

1/√2 of electrode width (null point)

Fig. 5. (a) Comsol simulation for the Electric field distribution above a pair of planar electrodes with voltage of +1V & -1V in two electrodes (160/40micron). (b) Four counter-rotating vortices are formed above the electrodes due to changes in tangential electric fields, which facilitate particles aggregation on electrodes.

3.5 Impedance analysis for optimization

The charging and ion migration process at electrodes can be represented by an equivalent circuit element. To analyze the effect of ACEO, it is necessary to develop an RC equivalent circuit for the pair of electrodes. Figure 6(a) is the equivalent circuit for the planar electrode in an aqueous environment. There are two paths that the planar interdigitated electrodes are connected as shown in the figure. One path goes through Ccell, which stands for the direct dielectric coupling between electrodes and it jumps the dielectric coupling through the fluid, and the environment. The other path goes through the fluid, which can be treated as resistance since it obeys Ohm's law. It is in series with capacitance for double layer charging at the interface of electrolyte and electrode on both ends. Figure 6(b) is the simplified RC equivalent circuit model. At low frequency, the reactance from double layer capacitance is high. Thereby a large portion of voltage drop happens within the double layer, suitable for the ACEO phenomenon to take place. At high frequency, time is limited for double layers to form, consequently inducing a great amount of surface charge in the double layer.

So, for the high frequency case the reactance gets smaller and more voltage drop across the resistive bulk fluid. As there is little voltage drop, hence little surface charge on the electrode, so ACEO becomes negligible at high frequency. By using the equivalent circuit model we can theoretically analyze the frequency range for the ACEO mechanism.

$$Z_{modeled} = \left[j\frac{1}{2*pi*f*C_{cell}} \quad || \quad \left(R_{sol} + j\frac{1}{2*pi*f*C_{tot\_dl}} \right) \right] + 2R_{lead}$$

The values of the above mentioned components in the equivalent circuit can be extracted by impedance measurements. 5 mV$_{rms}$ excitation level was used for impedance measurement. For 5 mV excitation level in can be very well assumed. In this condition faradaic charging will not take place, so we neglected the faradaic charging (Fig. 6b) for this analysis. The simplified equivalent circuit is shown in Fig 6(b). Following is the modeled impedance value,

$$Z_{modeled} = \left[j\frac{1}{2*pi*f*C_{cell}} \quad || \quad \left(R_{sol} + j\frac{1}{2*pi*f*C_{tot\_dl}} \right) \right] + 2R_{lead}$$

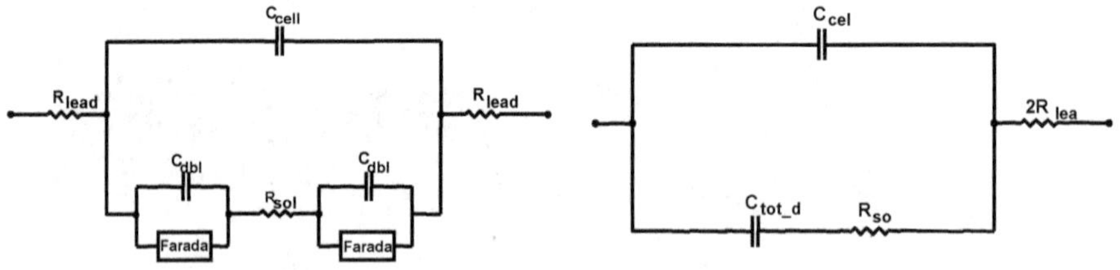

(a)   (b)

Fig. 6. Equivalent impedance for interdigitated pattern; (a) RC equivalent circuit for planar electrode configuration; (b) Simplified RC equivalent circuit for the modeling.

C_{cell} and C_{dbl} can be extracted from the experimental plot. At smaller frequency the C_{dbl} is dominated and at higher frequency $C_{cell/dielectric}$ dominants. Our extracted parameters are,

C_{cell} = 305 pF; C_{dbl} = 28.85 nF; R_{lead} = 18 Ω; R_{sol} = 966.234 Ω. We put all these values to our $Z_{modeled}$ and compare the plot with the experimental impedance plot from the impedance analyzer. The two plots matched fairly, so we can conclude that our extracted modeled parameter is correct.

The impedance measurements have been done for two different situations, 1) the electrode pair in DI water, and 2) the electrode pair in the DI water seeded with particles. We use information from impedance measurement to optimize the operating condition (signal frequency, magnitude etc.) of ACEO. Impedance plot for the two different excitation levels are also compared in figure 7. As seen from the figure, for high excitation voltage the impedance goes down. This is because the charge, Q is constant in the double layer, and the double layer capacitance is inversely proportional to the applied oscillator voltage (Q=CV). That is the reason the impedance plot for the two different excitation levels are different.

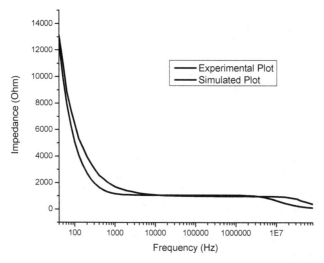

Fig. 7. Comparison of impedance plot between the experimental and the modeled data

Nyquist plot shows the frequency response for the linear system. Figure 8 shows the Nyquist plot for the equivalent circuit. The low frequency area of Nyquist plot denotes the double layer capacitive effect. Point "a" denotes the solution of resistance, which we extracted from figure 7 is 966.234 Ω. The graphical representation (figure 8) using a Nyquist plot represents the two parallel path of the microfluidic system, so that we can analyze the pre-dominant of the impedance in smaller and higher frequency range. The impedance analysis is very important to distinguish the effect of different electrokinetic forces.

From the figure 7 we also can see that the difference of the impedance with and without the particle is more in frequency below 1 KHz. So we adopted signal frequency range between 100 Hz to 1 kHz for particle trapping. The experiment was done at 500 mV$_{rms}$ oscillation level. Same characteristics were obtained at 1 V$_{rms}$. Our goal is to determine a higher velocity electrode pattern with the polystyrene particle, so both experiments and calculation were performed to determine the optimum signal magnitude for these four sizes of electrode pair (160/40, 160/20, 80/40, 80/20).

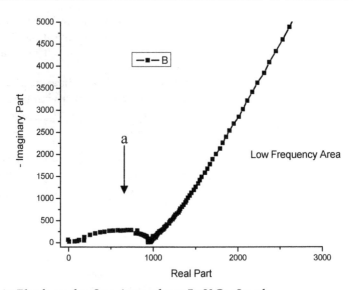

Fig. 8. Nyquist Plot for real vs Imaginary plot at 5mV Osc Level

4. Biased AC electroosmosis micropump

Micropump is critical to transport small amounts of fluid for many microfluidic applications, ranging from drug delivery, bio-fluid analysis, to microelectronics cooling. With the development of MEMS technology, micropumps have been designed and fabricated to integrate with lab-on-a-chip (LOC). Due to the large surface to volume ratio in microchannels, surface tension and viscous forces play an important role in the flow characteristics. So for micropumping action we need to choose the electroosmosis technique which benefits from the higher surface to volume ratio. For this reason electroosmotic action is suitable for miropumping action.

Electroosmotic (EO) pumping is the motion of bulk liquid caused by the application of an electric field to a channel. Electroosmotic (EO) pumps (*a.k.a.* electrokinetic pumps) have no moving parts and are capable of generating high flow rate per device volume. People use electroosmotic pumping to achieve both significant flow rates and pressures, and a fairly wide range of working electrolytes may be used (including deionized water, buffered aqueous electrolytes). These devices have significant pressure capacity in a compact structure. We have achieved flow rates in excess of 400 micron/second. EO pumps offer some advantages over other miniature pumps for microchannel cooling applications and integrated bio-analytical systems.

High-pressure capacity, millimeter-scale, porous-media based EO pumps have been demonstrated [32, 33], and most of the micropumps which are presented in the literature are DCEO micropumps. However, DCEO micropump suffers from high voltage operation (several kVs) and consequently excessive electrochemical reactions and electrolysis at the electrodes. This high voltage operation also creates the pH gradients and bubble which is not favorable for micropumping [28-32]. Again, many of the current fabrication techniques of porous-media EO pumps are not compatible with standard microfabrication processes and this poses a significant obstacle to the chip-level integration of EO pumps into microsystems. Our developed micropump is operated with smaller AC voltage and microfabrication compatible.

Reliability, compatibility, and cost are also criterions for selecting or designing micropumps. Micropumps must perform proposed functions without being damaged (re-usable), and at the same time must not bring changes in the medium. Compatibility with microsystems requires precisely pumping the desired range of fluid volumes and proper overall size of the micropump. Micropump can be used to manipulate the fluid volumes ranging from a few picoleters to hundreds of microliters for different biomedical applications, such as single molecule detection, species separation, antigen-antibody binding. The pump size is important for integrating the compact microsystem. The simplicity in design and fabrication of micropumps is also desirable. By using our fabricated micro-electrode array the problems of the electrokinetic micropumps can be solved.

4.1 The features of AC electroosmosis micropumps

Novel pumps based on ac electroosmosis are also investigated in this research. Here the AC electroosmosis is the main mechanism to drive fluids, which are quantitatively investigated by both experiments and Comsol simulations. Compared to other micropumps without moving parts, this AC electroosmotic micropump has following unique features.

1. Low operating voltage makes it superior in terms of device portability. In our design the AC electric fields are applied to pump fluids as small as 500mV bias voltage.
2. Avoids electrolysis and the resulting bubble generation. Currently EHD and MHD micropumps use DC fields to produce fluid motion, generating bubbles at the electrodes. In this research the applied AC fields has a frequency range of 100Hz to 5 KHz, allowing no time for bubble generation.
3. The dimension of our micropump is 5 mm long, 500 µm wide and 100 µm height. These pumps can be scaled down linearly to submicron/nanoscale or up by expanding laterally.
4. Minimizes pH gradient for using the smaller voltage.
5. The designed micropump is simple in structure. The channel and electrodes array are combined to achieve electroosmotic pumping. The designed ACEO micropump can also be integrated with lab-on-a-chip for miniaturization.

These features of the AC electroosmotic pump provide higher reliability, higher compatibility and lower cost. It can be possible for mass production of the electrode array on the wafer. The channel can also be fabricated using PDMS by using Si mold.

The designed AC electroosmotic pumps do not require open channels and hence are very useful for a sealed lab-on-a-chip system, where the fluid is circulating in closed channels. The use of this type of AC electroosmotic pump does not need open reservoir, avoiding undesired pressure due to different fluid heights in the reservoir.

4.2 Design of AC electroosmotic micropumps

Electroosmotic micropumps have been used widely in broad applications [33, 34]. Novel micropumps based on ACEO have been designed to drive fluids. AC electroosmotic micropumps can operate at lower voltage to avoid undesirable electrolysis and pH gradients. The main concept of getting the pumping action is to get the uni-directional flow by breaking the reflection-symmetry in the geometry or applied signal. Figure 9 shows the pumping action breaking the symmetry by changing the geometry [35]. By using the asymmetric pattern of electrode bigger vortices are generated on the larger electrode, which eventually dominants and produce uni-directional flow. The smaller electrode still produces the counter-rotating vortices that reduce the net flow.

Our design of AC electroosmotic micropump is based on the biased AC electroosmosis technique for symmetrical electrode array. The separation of the in-pair symmetrical electrodes (80 micron size) are 40 μm. We repeated the pair of electrodes to make the array of electrodes. The separation of the elctrode pair is 100 μm. For this particular configuration we name it 80/40/100 configuration. The corresponding numerical simulations using the finite element software FEMLAB are also conducted to verify the concept. The good agreement between the simulations and the experimental data regarding the uni-directional flow is also demonstrated.

4.2.1 Capacitive and faradaic charging effect

Biased ACEO is realized by applying biased AC signals over electrode pairs, leaving the electrolyte floating; therefore, two electrodes have different electrical potentials relative to the electrolyte. With a biased AC signal, $V_{applied}=V_0(1+\cos\omega t)$ over the electrodes, the left electrode is always positive and more prone to Faradaic charging, while the other is always negative and subject to capacitive charging. For the biased signal one electrode has a positive offset with the potential always greater than zero, while the other one lower than zero (Fig. 9). When the voltage exceeds the threshold for reaction, asymmetric vortices are formed above two electrodes as faradaic reactions take place at the positively biased electrodes. Faradaic reaction generates co-ions following Faraday's law. For the other electrode with a negative offset, counter ions are attracted to the electrode. Therefore, for two electrodes, same polarity of ions is induced. A unidirectional fluid loop is consequently formed by tangential fields, as shown in Figure 9. For the negative particles in an aqueous environment, they adhere to the stagnation line on the positively biased electrode. Henceforward, biased ACEO exhibits directional particle assembly.

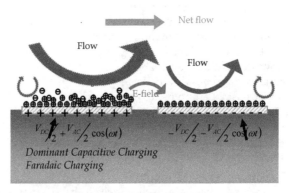

Fig. 9. Asymmetric polarization with appropriate magnitude can produce uni-directional micropumping.

Because most bioparticles are negatively charged, the DC bias can provide a synergy of AC and DC electrokinetics for more efficient particle collection. Electrophoretic/electrostatic force is exerted simultaneously with ACEO to move bioparticles towards positively-biased electrodes, as shown in Figure 9. Figure 9 already explained the electric field direction and net fluid flow for the asymmetric biased electrode pattern. Also in next section we have experimentally proved unidirectional particle flow. Breaking the charging symmetry on electrodes are main concept of producing uni-directional flow. With the biased AC signal, one electrode is at a higher voltage and hence undergoes Faradaic charging with cations, and the other electrode is at a lower voltage and hence is polarized by capacitive charging with the cations as well. This combination of the two polarization on the two electrodes of an electrode pair produces a uni-directional flow on each of the electrode pairs on electrode array. Which eventually produce a continuation of fluid flow and show the pumping action.

4.2.2 Unidirectional ACEO micropump

When using the asymmetric electrode pattern the direction of the pumping can not be reversed. For several biomedical applications there is a need for bi-directional flow directions. A common medical treatment procedure makes the use of a bidirectional flow control of one or more fluids to and from a patient. For these applications we have developed the biased AC electroosmotic micropump which can operate in both directions. Here we have broken the symmetry by applying asymmetric voltage on the symmetric electrode pattern, which eventually breaks the symmetry of the pattern. Figure 10 shows the electrode array, where L_{pair} is the separation between the two electrode pairs and L_{array} is the separation within the electrode pair. The mechanism of the designed pump is explained in Fig 10. As shown in the picture, both positive biased electrode (faradic charging) and the negative biased electrode (capacitive charging) generate the same positive charges on the electrode. The electrode surface has an excess of positive charges that creates a uni-directional flow in electrode pair, which eventually produce net flow on the array of electrodes. However, the coupling between neighboring electrode pairs will produce counterflows to the net flow produced within the pair and reduce the pumping efficiency. To reduce the undesirable coupling between the two adjacent pairs, L_{array} is kept larger than L_{pair}, but not so large that the flow loses its momentum.

Figure 10 is the schematic of pumping action for symmetrical electrode pattern with the biased applied voltage. The spacing Lpair and Larray will be the controlling factor for the pumping velocity. The rule of thumb is $L_{array} > L_{pair}$, so that the consecutive electrodes of Lpair and Larray do not form the EO flow in the reverse direction. From our earlier experiment we already have got L_{pair} of 20μm for the 80/20 configuration of our electrode pattern, which produces the highest fluid velocity.

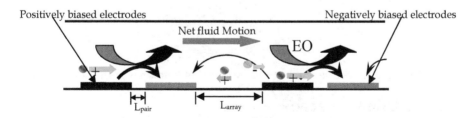

Fig. 10. Biased ACEO can produce uni-direction fuid motion, which also imparts differential velocities to particles with various charge/mass ratio.

Biased ACEO pump will be investigated by manipulating the faradic charging effect. We will also study the improvement in the pumping velocity. The pumping setup is shown in figure 11. The channel dimension is 350µm X 1.5mm X 5mm. The highest measured velocity for this pump is 150 µm/sec.

Fig. 11. Experimental setup of first version of the micropump.

4.3 Experimental results

ACEO flow is examined using microfabricated arrays of electrode pairs on silicon substrate. Au/Cr (90nm/10nm thickness) electrodes were fabricated by lift-off procedure in IC processing. Cr is the adhesion layer between the substrate and Au, and Au is in contact with electrolytes. The electrodes were 20 mm long, 0.1 µm think, 80 µm wide with a 20 µm separation (denoted as 80/20). Microfluidic chambers were formed by sealing silicone microchambers (PC8R-0.5, Grace Bio-Labs, Inc.) over the wafer, which have a height of 500 µm. Polystyrene spheres (1 µm diameter; Fluka Chemica) seeded in DI water was used to track fluid motion.

Figure 12 gives the comparative analysis of the microfluidic velocity generated using four types of patterned Au electrodes. The 80/20 configuration gives the highest velocity, which is suitable for the pumping application.

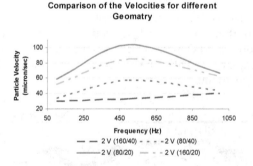

Fig. 12. Microfluidic Experiment result & comparison of four types of electrode geometry

4.4 Optimization of biased ACEO micropump

The goal of the research is to minimize the micropump reverse flow velocity. Obviously we increase the applied voltage it will increase the pumping velocity, and it will also increase the reverse pumping velocity (for the vortices). So, here we have focused on decreasing the reverse direction flow. Figure 13 explains the optimization concept by decreasing the channel height of the microfluidic chamber. Later on we also have presented the numerical simulation using Comsol Multiphysics for three different channel heights and verified the concept. The experimental result demonstrates that the thinner channel height ~ 100 μm increases the velocity of the micropump. In this case we also have demonstrated the bulk fluid flow, as here the surface to volume ratio is high.

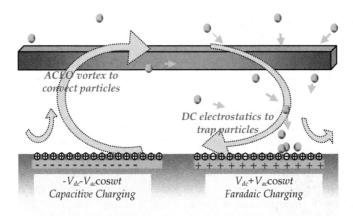

Fig. 13. Optimization Concept: to decrease the channel height to reduce the reverse flow

As seen from figure 13, the big vortex is suppressed / obstructed by the top wall of the channel. According to our analysis the vortex size is depended on the electrode geometry and the channel height. For the fixed electrode geometry the size of the vortex is only depended on the channel height, which is explained by Comsol Multiphysics simulation. The boundary condition for inflow and outflow was the key factor to run the simuation. Our findings show that smaller channel height will increase the surface flow, which will be described more in the next section.

Simulation of ACEO micropump has been performed prior to experiments. The simulation model, consists of two symmetrical electrodes of 80 μm width and the separation between the electrodes are 40 μm. The height of the chamber is 200 μm. From simulation we found that at the edge of the electrode the electric field is maximum. After simulating the electric field analysis the Navier-Stokes simulation is done to calculate the velocity field in the chamber. Initial fluid velocity is set to zero for the Navier-Stokes (NS) simulation. The fluid motions are generated by applied electric signals and a large vortex is formed for biased applied signal.

5. ACEO sensor integrated with microcantilever

We have investigated another microstructural design for ACEO devices, which is similar to a pair of parallel plates (Fig. 15). In this configuration, electrodes are facing each other, similar to a pair of parallel plates. The two face-to-face electrodes are asymmetric in design, so they produce non-uniform electric field. For mechanism identical to that of planar electrodes, surface EO flows are generated from the electrode edges inwards, and slow down to stagnation at the center, where particles are expected to trap.

5.1 ACEO particle trap

ACEO can transport the particles from a large region in the bulk fluid to the electrode surface. The flow velocity is important for optimizing the micropump and particle transportation. In contrast to electrophoretic and dielectrophoretic (DEP) velocity, which are typically limited to less than 20 microns per second [9], the ACEO velocity exceed 100 micron/sec. Figure 14(a) shows the initial distribution of particles when no signals are applied over the electrodes. Figure 14(b) shows that particles accumulated from both sides into lines at approximately $\frac{1}{\sqrt{2}}$ of electrode width [12]. This corroborates the theoretic prediction, since fluid velocity reduces at the null points of electric fields, and particles become trapped to the electrodes due to surface forces of between particles and electrodes.

For the biased ACEO experiments, applied voltage exceeds the threshold for reactions at V0=1.5V (i.e. high level & low level of biased voltage is 3V & 0V respectively). At the same voltage, the maximum flow velocity shifts to higher frequency compared with symmetric AC signals. This is because Faradaic polarization becomes suppressed at high frequency. Beyond 500Hz, microflows from capacitive charging are much stronger than those from Faradaic charging, so that the stagnation point on the left electrode disappears. At 100Hz, streamlines from capacitive charging and Faradaic charging become connected, forming a large vortex over the electrode pair and the particles aligned on the right electrode. Figure 14b also validates the null point formation by Comsol Multiphysics simulation in Figure5.

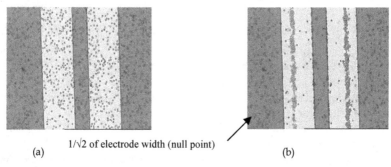

(a)   1/√2 of electrode width (null point)   (b)

Fig. 14. (a) Particle without the supply voltage (b) Experimental picture of the particles accumulating at the 1/√2 electrode width;

5.2 Electric field analysis of parallel plate particle trap

Most ACEO devices reported so far adopt a side-by-side (interdigitated) configuration. To integrate such design on microcantilevers would call for sophisticated microfabrication. We use a face-to-face configuration, very much like a pair of parallel plates, with one plate having smaller electrode area than the other, as shown in Figure 15. As the top and bottom electrodes are asymmetric, the tangential electric fields are generated which induces electro-osmotic fluid motion. For the first half cycle as in Figure 15a, the bottom electrode is at positive potential, and negative ions are induced at the interface of the electrode and fluid. These negative ions interact with the electric field and produce two counter-rotating vortices from electrode edges inward on the electrode surface, creating null point at the center of the electrode.

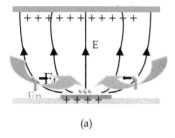 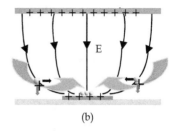

(a)   (b)

Fig. 15. Concept of Parallel plate particle trapping for an AC cycle; (a) during the half cycle when the bottom electrode has positive polarity and ITO coated top electrode is negative; (b) next half cycle with opposite polarity. Flow motion and induced charges is also shown

For the next half cycle as in Figure 15b, applied potentials switches polarity and the bottom electrode is at negative potential. Here the induced positive ions will interact with the electric field and again produces two counter-rotating vortices from electrode edges inward, and fluid motions are sustained thus the particles are trapped at the centre of the electrode. As the bottom plate is smaller than the top plate, the electric field is almost always normal to the top electrode, hence tangential force can be neglected.

The tangential electric field for the asymmetric electrode pattern induces electro-osmotic fluid motion in the bottom plate. It is the microfluidic flow that conveys particles from the bulk of the fluid onto the fluid surface. The stagnation point created at the centre of the bottom plate. The particles are trapped at the stagnation point of the fluid.

5.3 Microcantilever particle trapping using ACEO
This section explains the novel particle trapping method using microcantilever. Here we have presented the first integration of the microcantilever with the ACEO particle trapping mechanism. Recently microcantilever sensor technology has boomed and become a promising sensor technology. Microcantilever sensors have several advantages over many other sensor technologies, including faster response time, lower cost of fabrication, the ability to explore microenvironments, and improved portability. Cantilever resonance responses, such as frequency, deflection, Q-factor, and amplitude, undergo changes due to adsorption or changes in environment. Resonance frequency of a microcantilever can be used to detect particles. When the target is loaded on the microcantilever, the resonance frequency of microcantilever is going to change. That means for the mass loading on the cantilever the resonance frequency is supposed to go down.

The parallel plate design has been used to attract particles onto cantilevers for high sensitivity detection [14]. Because particle trapping/concentrating effect is more obvious with the smaller electrode, in our design the metal-coated cantilevers are facing one large electrode (covering a whole fluid chamber), so that particles will aggregate on the cantilevers. The tangential electric field of parallel plate configuration is generated for the asymmetric electrode pattern, which induces electro-osmosis fluid motion. In our design of microcantilever trap, the metal-coated cantilevers substitute the patterned bottom electrode, so that particles will aggregate on the cantilevers [15]. Figure 16 shows the experimental setup of cantilever particle trap.

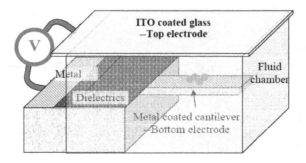

Fig. 16. Experimental Setup of Cantilever particle trap

As shown in Figure 16, photoresists (dielectrics) are coated on the conductive areas other than the cantilevers to suppress unwanted local EO flows. The ITO glass slide works as the top electrode, which is covering the whole fluid chamber. We have used Au-coated AFM probes as the MC, which has the dimension of 125μm X 30μm X 4μm. Tipped MC side was not used to avoid sharp electric fields.

Figure 17 shows the experimental results of trapping 200nm fluorescent particles on MC. After applying the AC signal (100Hz, 400mVp-p), suspended particles accumulate at the center of the cantilever from all directions. As time passes, more fluorescent particles from the surrounding area accumulated and formed bright object pattern. After the particle trapping on the surface, the MC was dried with AC signals applied, so that particles will not get dispersed by diffusion, surface tension, etc. Then the particle trapping effect was verified with AFM resonance measurement.

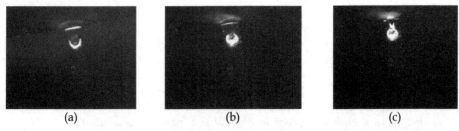

(a)   (b)   (c)

Fig. 17. Image sequence of 200 nm fluorescent particles trapped on the micro-cantilever;

5.3.1 Microcantilever particle trapping validation
To verify the concentrated particle trapping on MC, we also measure the resonance frequency of MC before and after trapping experiments. MC resonance frequency is inversely proportional to the differential mass of cantilever [21]. The sensitivity of a cantilever to mass loading is mainly determined by the excited cantilever resonance frequency.

$$\Delta m = \frac{K}{4\pi^2}\left(\frac{1}{f'^2} - \frac{1}{f^2}\right)$$

Where, Δm is the mass change;

K=sensor spring constant;

f = resonance frequency before mass adsorption;

And   f' is the resonance frequency during mass adsorption.

Changes in the mass and surface properties of the microcantilever through binding or hybridization of analytes to receptor molecules will directly influence its surface stress. This causes the microcantilever to deflect and the deflection is proportional to the analyte concentration and inversely proportional to mass loading. The more the particle concentration on ACEO-cantilever, the more is the bending. So the more mass on the cantilever means the lower resonance frequency. From our experimental result (Fig. 18) we have got the resonance frequency of the MC goes down to 276.07 KHz after particle trapped on the MC for ACEO, which translates to a change of mass. For a change in mass we get,

$$\frac{f_1}{f_2} = \frac{\sqrt{m_{eff} + \Delta m}}{\sqrt{m_{eff}}}$$

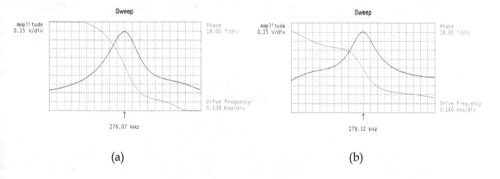

(a)   (b)

Fig. 18. Resonance frequencies of the MCs (measured with multimode AFM.)
(a) After the particle trapping by ACEO, 276.07 kHz, (b) Control experiment with no electric signal applied, 279.52 kHz.

Microcantilever dimension is 125μm x 30μm x 4μm. The volume of the MC is 1.5e-14 m³. Microcantilever is Si based, and the density of Si is 2330 kg/m³. So the mass of microcantilever is 3.495e-8 gm. By putting the frequency and mass values in equation (4.2), we have got an increase of 2.52% increase of mass for the frequency change from 279.52 KHz to 276.07 KHz. The change in resonant frequency as a function of the particle mass binding on the cantilever beam surface forms the basis of the particle detection scheme.

5.3.2 Analysis of particle trapping using SEM

Our research also shows that applying the electric field creates a certain crystal shape of the concentrated particles. Without applying the electric field, the particles are accumulating in layers, and formed no crystal shape (Fig. 19a). But when the electric field was applied for ACEO, the particles formed the close-packed layer of colloidal particles, and take the shape of crystalline structure, as shown in Figure 19(b). Compare with the Figure 19(a), the crystalline structure (Fig. 19b) increases the number of particles in the bottom layer [22].

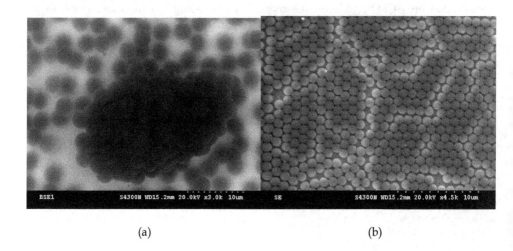

(a)   (b)

Fig. 19. SEM image of the particles; (a) Particles dried on the surface, no electric field is applied (b) For ACEO particle trapping, particles take the Crystal Shape.

6. Conclusion

AC electroosmosis, with its advantages of low power, low voltage and higher velocity, has been used to develop pumps. Such AC electrokinetic micropumps are presented in literature which have fast pumping (velocities » mm/s) velocity with low driving voltage of a few volts. But the literature review used the asymmetric pattern of electrodes. This chapter presents the novel micropump which utilizes the symmetric electrode pattern with the bias AC voltage. The chapter also focused on the MEMS microcantilever integration with ACEO.

The novel DC-biased, AC electroosmotic micropump operates in low voltage to avoid undesirable electrolysis and pH gradients. The developed uni-directional micropump prototype breaks the symmetry of the electrode array by applying the DC-biased AC signal, which has proven to be advantageous when compared to those pumps that use only asymmetric electrode patterning. In this work, a novel technique also has been developed to trap the particles and pump the fluid. Interfacing the micro-cantilever with ACEO mechanism has expanded its capability for biological, physical and chemical detection and makes the whole system ultra sensitive. The research work substantially enriches the portfolio of transducers, lab-on-a-chip (LOC) and MEMS that can be used in high performance miniaturized analytical systems.

7. Acknowledgment

This work is supported in part by the Nanotechnology for Undergraduate Education Program of the National Science Foundation (NSF-NUE) under Award Number EEC-1138205. The work is also supported from the internal fellowship grant from the University of Texas at Brownsville. We also want to thank Arizona State University (ASU) Nanofab Laboratory for assistance with microfabrication.

8. References

[1] Helene Andersson, Albert ven den Berg, " Microfluidic devices for cellomics : A review", Sensors and Actuators, B 92, (2003), 315-325.

[2] Vincent Studer, Anne Pépin, Yong Chen, and Armand Ajdari, "An integrated ac electrokinetic pump in a microfluidic loop for fast and tunable flow control", Analyst, vol. 129, pp. 944-949, 2004.

[3] C.-H. Chen and J.G. Santiago, "A planar electroosmotic micropump," J. Microelectromech. Syst., vol. 11, pp. 672-683, 2002.

[4] Lastochkin, D., Zhou, R., Wang, P., Ben, Y. and Chang, H.-C., "Electrokinetic Micropump and Micromixer Design Based on AC Faradaic Polarization", J. of Applied Physics , 96 , 1730 (2004).

[5] K. H. Bhatt, S. Grego and O. D. Velev, "An AC Electrokinetic Technique for Collection and Concentration of Particles and Cells on Patterned Electrodes," Langmuir 21, pp. 6603-6612, 2005.

[6] M. R. Brown, and C. D. Meinhart, "AC electroosmotic flow in a DNA concentrator," Microfluidics & Nanofluidics, ISSN:1613-4982 (Print) 1613-4990 (Online), 2006.

[7] Ronald F. Probstein, Physicochemical Hydrodynamics : An Introduction, 2nd Editon, New York : John Wiley and Sons, c 1994

[8] Stone, H.A. and S. Kim, "Microfluidics: Basic issues, applications, and challenges," AIChE J., 47(6), 1250-1253 (2001).

[9] A. Ramos, H. Morgan, N. G. Green and A. Castellanos, "AC electrokinetics: a review of forces in microelectrode structures", J. Phys. D: Appl. Phys. 31(1998) 2338–53

[10] A. Ramos, H. Morgan, N. G. Green and A. Castellanos, "AC electric-field-induced fluid flow in microelectrodes", J. Colloid Interface Sci. 217(1999) 420–2

[11] J. Wu, "Electrokinetic Microfluidics for On-Chip Bioparticle Processing," IEEE Trans. Nanotech., Mar. 2006.

[12] J. Wu, Y. Ben and H.-C. Chang, "Particle Detection by Micro- Electrical Impedance Spectroscopy with Asymmetric-Polarization AC Electroosmotic Trapping," Microfluidics & Nanofluidics, 1(2), pp. 161-167, 2005

[13] Julio M. Ottino and Stephen Wiggins, " Introduction: mixing in microfluidics", Phil. Trans. Royal Society Lond. A (2004) 362, 923-935.

[14] N. Islam, M. Lian, S. Swaminathan and J. Wu, "Micro/Nano- Particulate Fluid Manipulation in AC Electro-Kinetic Lab-on-a-Chip," 2nd ASM - IEEE EMBS Conf. Bio, Micro & Nanosyst., Jan. 15-18, 2006, San Francisco, CA, USA.

[15] J. Wu, N. Islam and M. Lian, "High Sensitivity Particle Detection By Biased AC Electro-Osmotic Trapping on Cantilever," 19th IEEE Int'l Conf. Micro Electro Mechanical Systems (MEMS 2006), Jan. 22-26, 2006, pp. 566 – 569, Istanbul, Turkey.

[16] N. Islam, M Lian, and J. Wu, "Enhancing Microcantilever Sensitivity with Integrated AC Electroosmotic Trapping," submitted to Bio-microfluidics Journal.

[17] J. Wu and H.-C. Chang, "Asymmetrically Biased AC Electrochemical Micropump," AIChE annual meeting 2004, Nov. 7 – 12, Austin, TX.

[18] Rosenthal A, Voldman J, "Dielectrophoretic traps for single-particle patterning," Biophysical Journal, 88(3), pp. 2193-2205, 2005.

[19] Meinhart C D, Wang D and Turner K, "Measurement of AC Electrokinetic Flows," J. Biomedical Microdevices, 5(2), 139-145, 2003

[20] Morgan H, Green N G, "Electrokinetics: Colloids and Nanoparticles," Research Studies Press Ltd: UK 2002

[21] Sepaniak M, Datskos P, Lavrik N, Tipple C, "Microcantilever Transducers: A New Approach in Sensor Technology," Analytical Chemistry, November 1, pp. 568A-575A, 2002.

[22] Green N G, Ramos A, Gonzalez A, Morgan H, and Castellanos A, "Fluid flow induced by nonuniform ac electric fields in electrolytes on microelectrodes. III. Observation of streamlines and numerical simulation," Physical Review E 66, 026305, 2002.

[23] Taylor, M. T., " Simulation of microfluidic pumping in a genomic DNA blood-processing cassette," Journal of Micromechanics and Microengineering, 2003. 13(2): pp 201-208.

[24] Laser, D. and J. Santiago, "A Review of micropumps," J. Micromechanics and Microengineering, 2004. 14:pp R34-R64.

[25] T.M. Squires and M.Z. Bazant, "Induced-charge Electro-osmosis," J. Fluid Mech., 509, pp. 217-252, 2004.

[26] Vincent Studer, Anne Pépin, Yong Chen, and Armand Ajdari, "An integrated ac electrokinetic pump in a microfluidic loop for fast and tunable flow control", The Analyst Journal, 129, 944-949, 2004.

[27] J. Wu, N. Islam and M. Lian, "High Sensitivity Particle Detection By Biased AC Electro-Osmotic Trapping on Cantilever," 19th IEEE Int'l Conf. Micro Electro Mechanical Systems (MEMS 2006), Jan. 22-26, 2006, pp. 566 – 569, Istanbul, Turkey.

[28] M. Z. Bazant, T. M. Squires, "Induced-charge Electrokinetic Phenomena: Theory and Microfluidic Applications", Phys. Rev. Letter, 92(6), 066101 (2004)

[29] A. Castellanos, A. Ramos, A. Gonzalez, N. G. Green, and H. Morgan, "Electrohydrodynamics and dielectrophoresis in microsystems: Scaling laws," J. Phys. D, Appl. Phys., vol. 36, pp. 2584–2597, 2003.

[30] Josh H. M. Lam, Raymond H. W. Lam, Kin Fong Lei, Winnie W. Y. Chow and Wen J. Li, "A Polymer-based Micro fluidic Mixing System Driven by Vortex Micropumps", Proceedings of the 5th World Congress on Intelligent Control and Automation, June 15-19, 2004.

[31] C. Liu, D. Cui, H. Cai, X. Chen, Z. Geng, "A rigid poly(dimethylsiloxane) sandwich electrophoresis microchip based on thin-casting method", Electrophoresis 2006, 27, 2917-2923

[32] N.G. Green, A. Ramos, A. Gonzalez, A. Castellanos, and H. Morgan, "Electrothermally induced fluid flow on microelectrodes," Journal of Electrostatics, 53 (2001) 71-87.

[33] Lastochkin, D., Zhou, R., Wang, P., Ben, Y. and Chang, H.-C., "Electrokinetic Micropump and Micromixer Design Based on AC Faradaic Polarization", J. of Applied Physics , 96 , 1730 (2004).

[34] Iki, N., H. Hoshino, and T. Yotsuyanagi, " A capillary electrophoretic reactor with an electroosmosis control method for measurement of dissociation kinetics of metal complexes," Analytical Chemistry, 2000. 72(20): pp. 4812-4820.

[35] A. Ramos, A. Gonzalez, A. Castellanos, N.G. Green, and H. Morgan, "Pumping of liquids with ac voltages applied to asymmetric pairs of microelectrodes," Phys. Rev. E, 67, 056302 (2003).

[36] Shaorong Liu, Qiaosheng Pu, and Joann J. Lu, "Electric field-decoupled electroosmotic pump for microfluidic devices", Journal of Chromatography A, 1013 (2003), 57-64.

[37] Ping Wang, Zilin Chen, and Hsueh-Chia Chang, "A new electro-osmotic pump based on silica monoliths", ELSEVIER-Sensors and Actuators B113 (2006), 500-509.

[38] A. Ramos, H. Morgan, N. G. Green, A. Castellanos. "AC electrokinetics: a review of forces in microelectrode structures". J. Phys. D: Appl. Phys. Vol 31. pp 2338-2353.

[39] N.G. Green, A. Ramos, A. Gonzalez, H. Morgan, and A. Castellanos, "Fluid flow induced by nonuniform ac electric fields in electrolytes on microelectrodes. III. Observation of streamlines and numerical simulation," Physical Review E, 2002, Vol.66, 026305

[40] Shuhuai Yao and Juan G. Santiago, "Porous glass electroosmotic pumps: theory", ELSEVIER- Journal of Colloid and Interface Science, 268 (2003), 133-142.

[41] Chuan-Hua Chen and Juan G. Santiago, "A Planar Electroosmotic Micropump", Journal of Microelectromechanical Systems, Vol.11, No.6, December 2002.

[42] P. H. Paul, D.W. Arnold, and D. J. Rakestraw, "Electrokinetic generation of high pressures using porous microstructures," in _-TAS 98, Banff, Canada, 1998.

[43] S. Zeng, C. H. Chen, J. C. Mikkelsen Jr., and J. G. Santiago, "Fabrication and characterization of electroosmotic micropumps," Sensors Actuat. B, vol. 79, pp. 107-114, 2001

[44] A. Ramos, A. Gonzalez, A. Castellanos, N.G. Green, and H. Morgan, "Pumping of liquids with ac voltages applied to asymmetric pairs of microelectrodes," Phys. Rev. E, 67, 056302 (2003).

[45] N. Islam and J. Wu, "Microfluidic Transport by AC Electroosmosis", Journal of Physics: Conference Series, 34, pp. 356 – 361, 2006s

Part 2

MEMS Characterization and Micromachining

5

MEMS Characterization Based on Optical Measuring Methods

Tong Guo[1], Long Ma[2] and Yan Bian[3]
[1]Tianjin University
[2]Civil Aviation University of China
[3]Tianjin University of Technology and Education
P.R. China

1. Introduction

Micro Electro Mechanical Systems (MEMS) is developed based on the semi-conductor technology, however, relative material, design, fabrication, simulation, packaging and test are more complex than those in semi-conductor technology. In the primary stage, MEMS technology focused on the design and development, now on the commercialization and improving reliability and decreasing cost and price. So test is increasingly important to MEMS technology and testing cost is about 1/3 of the whole cost of MEMS. In order to improve the production and decrease the cost, producers and researchers pay more attention to MEMS test to solve all the testing problems from design to packaging process.

There are a number of methods to carry out these measurements, such as scanning electron microscopy (SEM), atomic force microscopy (AFM), stylus profiler, and optical profiler, etc. Every method has its advantages and disadvantages.

- SEM is one of the most common measurement tools. However, nearly all nonconductive specimens examined using SEM need to be coated with a thin film of conducting material. This may result in bending or distortion of the device, especially where free structures such as cantilever beams. SEM tests are also time consuming and not suitable for a production environment.

- AFM has been suggested as a MEMS measurement tool. As with SEM, analysis may be slow (about 20 min/device), and the limited measurement range of an AFM (100 µm×100 µm×5 µm, Veeco multimode AFM) means that it is unable to investigate large samples or out-of-plane devices such as the cantilevers. It is also difficult to examine packaged devices using an AFM.

- Mechanical stylus surface profilers are commonly used for dimensional measurements in MEMS. While these can measure step heights with a high accuracy, they are not suitable for the analysis of freestanding structures, where the stylus may break the device under test. Deep, high aspect ratio devices also pose problems, as the stylus probe may be too large to accurately reproduce the surface profile.

If MEMS devices need to fit the large-scale production, it is essential that these measurements must be cheaply and easily made at the wafer level, without the need for large space, expensive packaging or destructive test methods. Optical techniques can offer solutions to many of these problems. This chapter uses computer micro-vision and microscopic interferometry to carry out MEMS measurements, including dimensional (static) and moving (dynamic) properties analysis. The moving properties can be classified into in-plane (lateral) movements and out-of-plane (vertical) movements. The techniques involved are simple, fast, non-destructive, requiring no sample preparation and may be carried out at wafer level - all important requirements for high volume production.

2. System set-up

2.1 Microscopic interferometer
The interferometer is the device which can generate the interferogram patterns. The typical microscopic interferometers used in MEMS measurement include Michelson-type, Mirau-type and Linnik-type. The scheme of optical structures in these interferometers is shown in figure 1.

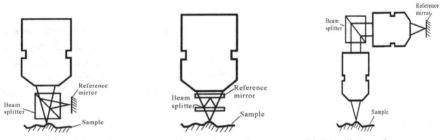

(a) Michelson interferometer   (b) Mirau interferometer   (c) Linnik interferometer

Fig. 1. Schematic layout of the three types of interferometers

Inside the interferometer the light is divided into two beams, one beam is guided on the sample surface as the test beam, and the other one goes to the reference mirror as the reference beam. When the OPD is within the coherence length, the reflected test beam and reference beam will interfere with each other and produce interference fringe patterns. Figure 1(a) presents the Michelson interferometer, where a beam splitter is placed in front of the objective. Hence, the working distance of the objective must be relatively large which makes the magnification of Michelson interferometer the lowest of the three types of interferometers shown in figure 1. Figure 1 (b) is the Mirau interferometer, which is widely equipped in lots of commercial optical profilers. Inside this interferometer, the light is split by a very thin and flat optical mirror. As result, the mechanical structure of this interferometer is more compact than that of Michelson type. Besides, the light path between the test beam and the reference beam is very similar, which can also minimize optical disturbance. Figure 1 (c) illustrates the layout of the Linnik interferometer, from where one can see the beam splitter is located behind the objective. Therefore, Linnik interferometer can use the objective with shorter working distance and higher magnification. However, the unconformity between the two objectives may cause measurement errors, and that is why in the application of Linnik interferometer the two objectives must be matched very well.

2.2 Measurement system

Two sets of measurement systems were developed in our laboratory. The first one is for static characterization, the other one is for dynamic characterization.

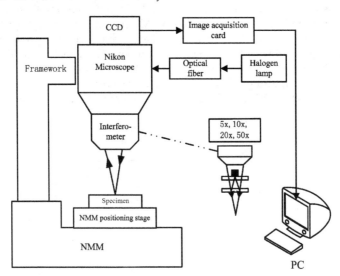

Fig. 2. Scheme of the static characterization system

As a foundation of our system, the nano-measuring machine (NMM) was used, which was developed in Ilmenau University of Technology of Germany for surface measurement within a measuring volume of 25mm×25mm×5mm. Three commercial interferometers inside the NMM read the stage position in real time so that the positioning control loop can assure a movement resolution of 0.1 nm. The sensor arrangement provides an Abbe error-free measurement on all three NMM axes. Based on the NMM, various measuring systems could be integrated.

The layout of the first experimental system used in this context is illustrated in figure 2. To increase the optical system robustness, we chose a commercial interferometer system with halogen lamp fibre illumination from NIKON. A CCD camera recorded the interferogram at equidistant positions and transfered them into a computer for data processing. In the laboratory, the system was assembled inside a cover where a vibration isolation table was provided. Additionally, in order to further decrease the influence of vibrations, we placed the entire equipment on a special independent foundation.

The second system is called micro motion analyzer (MMA). The structure of MMA is shown in figure 3. MMA was equipped with several long working distance optical objectives from Zeiss company, including 5×, 10×, 20× and 50×. There were two kinds of light sources: one was a high performance LED (Nichia NSPG 500S LED, central wavelength was 525nm, optical bandwidth was 40nm) to be used in the in-plane motion measurement; the other was a LD (Hitachi HL6501MG, the power was 50mW, the wavelength was 658nm) to be used in the out-of-plane motion measurement. A nano-positioner (PI P-721.CL, capacitor sensor feedback control in close loop) was used to shift the phase of the interferogram and adjust the optical path with sub-nanometer resolution. The bright field and interference field were switched by putting a stop plate in the reference optical path. Images were captured by a digital CCD camera (Microvision, 1280×1024 pixels, 9μm pixel distance, 100% filling factor, 10 bits grayscale resolution), then transferred to the server to be post-processed. The user can transfer and receive instructions and data through TCP/IP protocol on a client PC. The system was also equipped with a signal generating unit and a high voltage amplifying unit to output high voltage signal and stimulate the tested MEMS device. The system was placed on a vibration isolation table (TMC Lab Table) to decrease the effect of outside vibration. The system was also equipped with a probe station (Karl Suss, Germany) to test the unpackaged MEMS devices. Here digital phase lock loop (PLL) made the stimulating signal, illuminating signal and the image capturing synchronized. The frequency range of Arbitrary Waveform Generator (AWG) was from 1 Hz to 10 MHz, and the sampling frequency was 40 MHz.

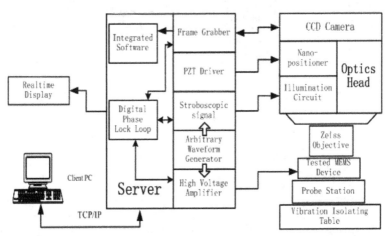

Fig. 3. Structure diagram of the MMA

3. Measurement principles

3.1 Static characterization

Static characterization is a very important aspect in MEMS technology, especially for the design and manufacture of micro- or nano-sensors and actuators. In general case, the goal of static characterization is to derive the geometrical parameters, such as surface topography, profile, waviness, roughness and depth-width ratio, etc.

3.1.1 Phase-shifting interferometry (PSI)

If we denote the coordinates in the plane of the CCD array as (x, y), we can write the intensity of the interference pattern I(x, y) as:

$$I(x,y) = A(x,y) + B(x,y)\cos[\phi(x,y) + \varphi] \qquad (1)$$

Where $A(x, y)$ is the function representing the mean background intensity of the interferogram; $B(x, y)$ is the function determining the modulation of the interferogram; $\phi(x, y)$ is the function dependent on the unwrapped phase $\theta(x, y)$ which represents the height change on the sample; phase φ is a constant, which can be adjusted by changing the position in the axial direction between the objective and the sample.

Using a series of I measured at different phase φ value, Hariharan phase-shifting algorithm is used to extract height information on devices' surface in I. The intensities are I_1, I_2, I_3, I_4 and I_5 respectively. The formula of calculating wrapped phase ϕ is as follows.

$$\phi(x,y) = \arctan\left[\frac{2(I_2 - I_4)}{2I_3 - I_1 - I_5}\right] \qquad (2)$$

This algorithm need change $\pi/2$ between two successive image capturing. This algorithm is repressive to the calibration error of the phase-shifter and the nonlinear errors of the detector, so it improves the measurement accuracy.

Measurements using single-wavelength interferometry have an inherent 2π phase ambiguity, which means phase maps must be unwrapped by removing artificial 2π jumps before they can be properly interpreted in terms of surface height. This process is called phase unwrapping. Now there are many kinds of phase unwrapping algorithms. In the experiments, considering about the testing speed and the real quality of the interferogram, quality-guided path following algorithm is used to retrieve the actual phase map. The surface height can be calculated by the following formula.

$$h = \frac{\theta}{4\pi}\lambda \qquad (3)$$

Here, λ is the wavelength of the light source.

Figure 4 shows the wrapped phase map and the unwrapped map in the process of dealing with the captured images. The tested sample is a membrane.

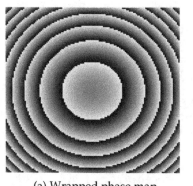

      (a) Wrapped phase map   (b) Unwrapped phase map

Fig. 4. Pictures of processed results

3.1.2 White light scanning interferometry (WLI)

PSI is widely used to test smooth surfaces and is very accurate, resulting in vertical measurements with sub-nanometer resolution. However, PSI cannot obtain a correct profile for objects that have large step changes because it becomes ineffective as height discontinuities of adjacent pixels exceed one quarter of the used wavelength ($\lambda/4$), which is also called "phase ambiguity".

White light scanning interferometry (WLI) provides a good solution to overcome "phase ambiguity". Here we track point P and Q in figure 5 to illustrate the scan process of WLI. When Q stays outside the coherence length of the light source, the intensity detected is nearly the same with the background irradiance, corresponding to t_1 to t_2. Afterwards, from time t_2 when Q starts to move into the coherence length zone, the signals extracted of point Q will be kept modulating by the interferogram until it travels out at t_5. During the scan, the maximum visibility will occur at t_4 where the testing beam matches the reference beam. The above process is roughly similarity for point P. Once retrieved the correlogram, the peak position of the intensities can then be used as an indicator of the surface relative height.

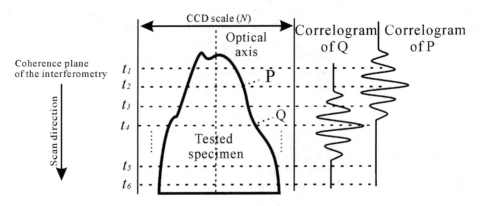

Fig. 5. Schematic diagram of white light vertical scanning interferometry

The white light correlogram recorded by the CCD camera in the white light interferometry can be expressed as:

$$I = I_0 \left[1 + \gamma g(z) \cos(\phi + \alpha) \right] \tag{4}$$

Where I is the measured intensity, I_0 is the background irradiance value, γ is a constant, $g(z)$ is the fringe visibility in the form of Gaussian function, ϕ is the phase value depended on the optical path difference and the α is the additional phase term due to different reflections. Generally, the maximum location of $g(z)$ is usually extracted for the height evaluation. This process can be done either in spatial space or in frequency domain.

So far, the white light signal demodulation techniques can be basically categorized into two main groups: spatial domain algorithms and frequency domain algorithms. The first group consists of polynomial interpolation, centroid method and Hilbert transform, whereas the latter has Fourier transform, wavelet transform and spatial frequency domain analysis, etc. In general case, algorithms from the second group can perform the surface characterization with a higher resolution. Nevertheless, because of the transform procedures from the spatial domain to frequency domain (various types of convolutions), the second group algorithms are usually time consuming; on the contrary, the first group is efficient but with lower resolution. The other methods are generally based upon the above algorithms and will not be mentioned here.

3.1.3 White light phase-shifting interferometry (WLPSI)

3.1.3.1 Phase extraction

Carré proposed a new phase shifting interferometry in 1966. Unlike the conventional PSI, Carré's method does not require several fixed phase steps (for example, $\lambda/8$), but only equal ones, which makes it much easier for most PZTs to fulfill. Suppose the phase step is set to be 2δ, Carré's method can be expressed as:

$$I_1(x,y) = I_0\{1 + \gamma \cos[\phi(x,y) - 3\delta]\} \tag{5}$$

$$I_2(x,y) = I_0\{1 + \gamma \cos[\phi(x,y) - \delta]\} \tag{6}$$

$$I_3(x,y) = I_0\{1 + \gamma \cos[\phi(x,y) + \delta]\} \tag{7}$$

$$I_4(x,y) = I_0\{1 + \gamma \cos[\phi(x,y) + 3\delta]\} \tag{8}$$

Where I_1 to I_4 are the recorded intensities, I_0 is the background irradiance, γ is a constant indicated the visibility and ϕ is the phase term needed to be extracted.
Then the phase term ϕ can be computed as:

$$\phi = tg^{-1} \frac{\sqrt{[(I_1 - I_4) + (I_2 - I_3)][3(I_2 - I_3) - (I_1 - I_4)]}}{\left|[(I_2 + I_3) - (I_1 + I_4)]\right|} \tag{9}$$

Here ϕ falls within $(-\pi, +\pi]$ and the surface information can be written as:

$$h = \frac{\bar{\phi}}{4\pi} \lambda \tag{10}$$

Where h is the relative height of the surface, $\bar{\phi}$ is the unwrapped phase term, λ is the wavelength used. The relative heights on every point together can then give a surface 3D map of the tested sample.

However, in the white light correlogram, the existence of $g(z)$ will make g be a function of the spatial position along the scanning direction, which means error occurs when we employ the Carré method to carry out the phase computation from a white light interferogram. In the next, we will use computer simulation to study how much this kind of error affects the measuring accuracy.
The generated discrete white light signal $I(n)$ is as follows:

$$I(n) = 200 + 200\exp\{-\frac{[(n-20) \times \Delta]^2}{\sigma^2}\} \cos[\frac{4\pi}{\lambda}(n - 20) \times \Delta + \frac{\pi}{7}] \tag{11}$$

Where Δ is the scanning step 60 nm, σ is set to be 500 nm($\sigma = l_c/2\pi$, l_c is the coherence length of the light source), the additional phase term due to different reflections is set to be $\pi/7$. Figure 6 is the phase computation simulation, which indicates that: phase extraction error owing to the visibility variety is minimum at the position of the coherence peak; on the other hand, the phase error has a trend of becoming bigger while the phase position travels away from the centre. The minimum phase error is less than 0.02 rad and the equivalent measurement error is smaller than $\lambda/500$. Assumed the wavelength is 600 nm, this error could be about 1nm and is negligible in comparison with the environment disturbance. This can be explained as the weak modulation effect of the Gaussian function around the zero order fringe of the white light signal, where the value of the Gaussian function can be treated as 1(for normalized Gaussian function).

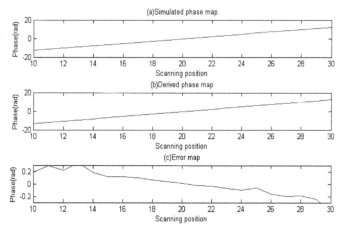

Fig. 6. Phase extraction simulation

3.1.3.2 Height evaluation

Based on the phase computation simulation, the relative height of the tested profile h can then be determined as:

$$h = (stepnumber - peakstep) \times \Delta - t \times [\frac{(\phi + 2k\pi)}{4\pi} \lambda]$$ (12)

Where *stepnumber* is the scanning numbers, *peakstep* corresponds to the phase term which comes out of Carré method, Δ is the scanning step and t is the NA parameter. Concerning the NA parameter, both Katherine Creath and C.J.R.Sheppard have given lots of detailed research on the relationship between the objective NA and the fringe width. Here we take this relationship into account and use the results from Ingelstam's equation to calculate this parameter.

$$t = 1 + \frac{(NA_{eff})^2}{4}$$ (13)

Where NA_{eff} is the effective numerical aperture.

3.2 Dynamic characterization

The MMA is a highly integrated video microscope, using stroboscopic techniques to capture images of small, fast moving targets. The MMA uses both bright field and interference field based illumination modes combined with sophisticated machine vision algorithms to quantify micro motions. The MMA server and optics head combine the video microscopy with interferometry. However, successful measurements demand several characteristics in the target. The MMA needs two things: a target that can be moved in a periodic manner, and a target whose video image has contrast or structure when illuminated.

3.2.1 In-plane motion measurement using video microscopy

The MMA calculates the motion of a selected region in a sequence of images using machine vision algorithms. The MMA algorithms are hybrids constructed from a broad class of algorithms (gradient based) that exploit changes in brightness (grayscale values) between images and are capable of measuring motions smaller than the individual pixel size.

Fast moving targets appear blurry. One way of looking at fast moving targets is to slow down their apparent motion using a strobed light source. The scheme shown in figure 7 outlines how carefully timed pulses of light can capture snapshots (samples) of an object motion. These snapshots or images can then be used to reconstruct the displacement trajectory of the target. This is the case for displacements in the focal plane of the microscope or for displacements along the optical axis. It turns out that sometimes different types of images are better for measuring different types of displacements. The MMA optics module provides illumination for two types of images, both of which are acquired in a similar manner.

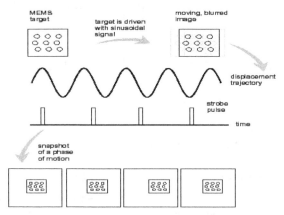

Fig. 7. Freezing rapid movement by stroboscopic illumination

3.2.1.1 Bright field image

By shining a monochromatic incoherent beam of light through the objective lens and uniformily illuminating the target, a bright field image is produced. This is the familiar view through the microscope eyepiece. A bright field image of a MEMS device is shown in figure 8(a).

Optical flow is used to describe the measured motion of brightness patterns between images. The algorithms used by the MMA are optical flow algorithms and are based on two important assumptions or constraints: 1) The brightness of a target region is constant over time, 2) The target region moves as a rigid body.

Motion of the target modulates brightness in the image. The image of a MEMS target moves across an array of pixels. The position of the image is shown at two consecutive timesteps and the registers at the bottom of each image represent the nominal brightness of the corresponding pixel in row 6 (in figure 9). Even sub-pixel motion has measureably changed the brightness in the pixels.

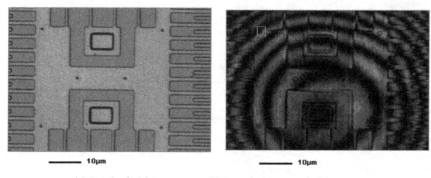

(a) Bright field image   (b) Interferometric field image

Fig. 8. Images of the tested micro-resonator

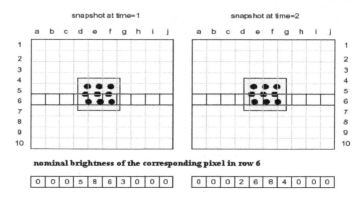

Fig. 9. In-plane motion measurement using optical flow algorithm

The optical flow algorithms used by the MMA work for in-plane or out-of-plane measurements. For out-of-plane measurements, image sequences can be collected at different planes of focus (optical sectioning). In this way, brightness gradients can be sampled along the optical axis and then used in a manner identical to in-plane sequences.

3.2.1.2 Interference field image

Figure 8(b) is the interference field image, which is formed by the reflected light from the tested sample and the reference mirror in the optics module. The interferogram in the picture is sensitive to the optical path change of the sample, which can be used to calculate the height of the sample and the out-of-plane motion between images.

3.2.2 Sub-pixel displacements measurement

The final result of the computation is an extremely powerful tool capable of measuring motions of magnified targets well below the resolution of human vision. Although, the resolution of a single static image is limited in a theoretical sense by the wavelength of the light used to generate the image, the resolution of a motion measurement is limited by the sensitivity of the CCD camera as shown in figure 9. Sub-pixel motion of an object or region (larger than a pixel) can be measured by using the modulation of the brightness of relevant pixels. It is also clear from figure 9 that changes in the measured brightness can be produced by factors other than motion of the target such as noise in the CCD camera's electronics or fluctuations in the illumination intensity.

3.2.3 Qualifying displacements

In order to quantify in-plane or out-of-plane displacements, the algorithms need a length scale. It means that different measurements need corresponding length scale.

For bright field images (in-plane measurements and optical sectioning), an in-plane length standard such as a grating (a target with a ruling of known spacing) is used to calibrate the length per pixel for a given magnification. The algorithms then convert the calculated displacements from pixels to micrometers.

Interference images have a more convenient length scale available. In this case, measurements are quantified using the known wavelength of the illumination source. This length standard is independent of magnification.

3.2.4 Noise floor analysis

After the data acquisition process is finished, the stimulating system is turned off and the measurement is done again. These two kinds of data are analyzed using the same processing method. Then repeating it five times, the noise floor is described by the RMS value.

4. Experimental results

4.1 Step structure measurement
4.1.1 10 µm standard step height measurement

A 10 µm standard step height fabricated by VLSI (9.976 µm+0.028 µm) was measured, as shown in figure 10, the mean height is 9.984 µm, while the standard deviation is 0.010 µm. The result comparison was presented in figure 11. We can clearly see the step height derived from WLPSI stays in the middle of all the results. Since the algorithms from the second group can achieve higher measurement resolution, the WLPSI can improve the measurement resolution compared with that of the algorithms from the first group.

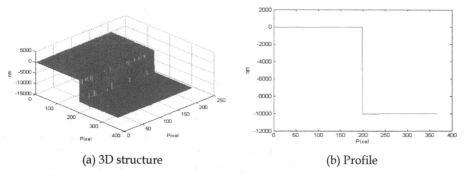

(a) 3D structure   (b) Profile

Fig. 10. Measurement of the 10 µm standard step height

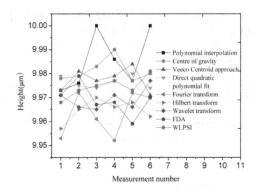

Fig. 11. 10 µm standard step height measurement comparison

4.1.2 44 nm standard step height measurement

Many coherence peak detecting algorithms in white light interferometry perform the height evaluations through either recovering the envelop function or locating the centroid of the correlogram. It works quite well for most surfaces, however, when it comes to some step height alike structures that lower than the coherence length of the illuminator, overshoot will be observed at these discontinuities. This overshoot is also known as batwings, which comes out in form of the high frequency information in the measurement results. Normally, the larger magnification objective always has the higher cutoff spatial frequency, so it will also bring stronger batwings. The batwings is shown in figure 12, where a 44 nm standard step height manufactured by VLSI (43.2 nm±0.6 nm) was measured by Fourier transform.

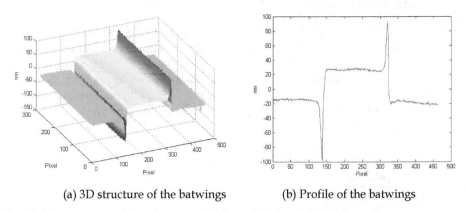

(a) 3D structure of the batwings   (b) Profile of the batwings

Fig. 12. Measurement of the 44 nm standard step height with batwings

During the vertical scanning, on condition that the step discontinuities is lower than the coherence length, the light diffracted by the top edge and the light reflected from the bottom will interfere with each other and then travel back to the interferometer. It is because of this diffraction that deforms the correlogram and produces batwings. However, the batwings never occurred in PSI measurements, which gives us a clue to find out the solution for this problem: intensities are much easier to be affected than the phase information in the white light signals. In this work, we repeated the measurements with WLPSI on the above 44 nm standard step height in figure 12. As illustrated in figure 13, it gave out correct results without batwings.

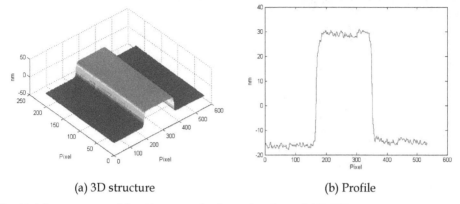

(a) 3D structure   (b) Profile

Fig. 13. Measurement of the 44 nm standard step height with WLPSI

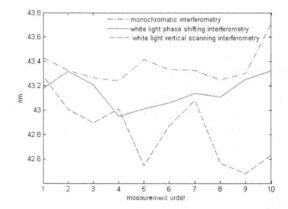

Fig. 14. Comparison of the results from the 44 nm standard step height calibrations

	PSI	WLPSI	WLI
Mean height(nm)	43.36	43.16	42.84
Standard deviation(nm)	0.14	0.13	0.27

Table 1. Measurement results of the 44 nm standard step height using different methods

Figure 14 and table 1 show the 44 nm standard step height evaluations comparison between PSI, WLPSI and WLI. We can clearly see that the results from WLPSI ended up basically in the middle of these three methods and it is close to the results from PSI, which also shows the ability of WLPSI in improving the measurement accuracy compared with the traditional WLI.

4.2 Micro pressure sensor measurement

Experiments were done on a micro pressure sensor using PSI under the air condition. From figure 15(a), a deformed membrane can be seen to show the air pressure. The deform value shows the change of the pressure. From figure 15(b), the height difference between the highest position and the lowest position is 1345.7 nm, which is matched with the designed value.

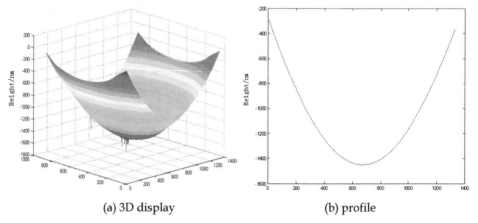

(a) 3D display   (b) profile

Fig. 15. Results of micro pressure sensor measurement

4.3 Film structure measurement

Combining with Otsu method from image segmentation technique, we measured a film thickness standard in WLI measurement (centre average thickness: 1052.2 nm±0.9 nm, refraction index: 1.46, model number: FTS4-10100, VLSI) which was calibrated by an ellipsometer in 632.8 nm wavelength. The system equipped 10× Mirau objective to perform a set of repeated measurements". The thickness and the surface topography were successfully extracted, which were shown in table 2 and figure 16. The R_a values of the upper and lower surfaces were 7.30 nm and 7.32 nm, respectively.

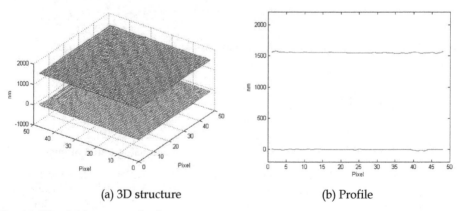

(a) 3D structure   (b) Profile

Fig. 16. Film thickness standard measurement

No.	Thickness(nm)	No.	Thickness(nm)	Mean value(nm)	Standard deviation(nm)
1	1053.1	6	1050.9		
2	1053.1	7	1052.3		
3	1050.0	8	1052.1	1051.8	1.43
4	1048.8	9	1052.1		
5	1050.9	10	1053.1		

Table 2. Measurements on film thickness standard

4.4 Micro-resonator measurement

With WLPSI, a micro resonator manufactured by Microelectronics Center of North Carolina (MCNC) was measured. With a 20× Mirau objective, the micro resonator was illuminated by a white light illuminator (central wavelength is 600 nm), the scanning range was configured to be 9 μm and the scanning step was 45 nm. The result is shown in figure 17. The height of the comb-finger profile is approximately 3.8 μm.

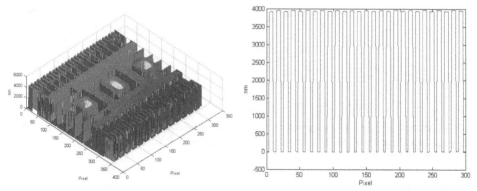

(a) 3D structure of the resonator   (b) Comb-finger profile of the resonator

Fig. 17. Measurement of the resonator

4.5 Dynamic characterization
4.5.1 In-plane motion measurement

A 20× objective was used in the dynamic experiments. The stimulating signal was sine waveform with 10V amplitude, 20V offset voltage. The exposing time of CCD was 20ms for LED illumination and 100μs for LD illumination. The strobed pulse percent was 100%. The strobed phase number was 8 for in-plane motion measurement and 16 for out-of-plane motion measurement.

In order to get the resonant frequency of the tested device, the sweep-frequency measurement in a large range was firstly needed. The frequency increases in the logarithm mode, so the range of resonant frequency can be decided through one time frequency sweeping measurement. Then another frequency sweeping measurement can be done in a small range. In summary, smaller range can achieve more times measurement, and the gotten resonant frequency will be more accurate. Figure 18 shows the experimental results in the range from 100 Hz to 100 kHz, the sweeping number is 21. After that, the frequency sweeping range can be decreased to 22 kHz~25 kHz, the sweeping number was 100 (see in figure 19). From that, the resonant frequency of the microstructure can be obtained. From the amplitude-frequency and phase-frequency curves, the device can be fitted into a second order system, the resonant frequency of the device is about 23.41 kHz, the maximum moving amplitude is around 764.52 nm, so the quality factor can be calculated as Q=23.41/(24.04-22.87)=20. Because of the air damp during the moving process of the device, the quality factor is not high. All data will be fed back to the designer and the quality factor of the device can be increased through improving the structure of the micro-resonator. Then five times of experiments without stimulating signal were done. The noise floor was calculated as 0.56 nm.

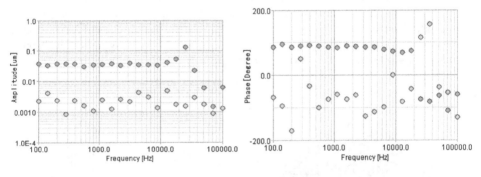

(a) Amplitude-frequency curve   (b) Phase-frequency curve

Fig. 18. Large range sweep-frequency measurement of in-plane motion. The blue curve is the result in x direction. The yellow curve is the result in y direction.

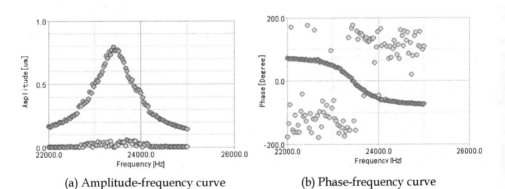

(a) Amplitude-frequency curve   (b) Phase-frequency curve

Fig. 19. Small range sweep-frequency measurement of in-plane motion. The blue curve is the result in x direction. The yellow curve is the result in y direction.

In order to study the relationship between the stimulating voltage and the movement, amplitude sweeping experiments were performed. The stimulating frequency was 23 kHz with the starting amplitude 0% and the ending amplitude 100% (10 V). The sweeping number was 21 and the amplitude increased linearly. The experimental results were shown in figure 20. The curves show that the relationship between the moving amplitude and the stimulating voltage is linear which is consistent with the computer simulated results. It draws the conclusion that the design and fabrication process of the device are valid and the device has good dynamic behaviors.

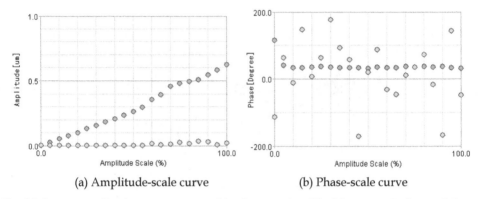

(a) Amplitude-scale curve   (b) Phase-scale curve

Fig. 20. Sweep-amplitude measurement of in-plane motion. The blue curve is the result in x direction. The yellow curve is the result in y direction.

4.5.2 Dynamic profile measurement
A region of interest (ROI, see in figure 21) on the device was tested to study the dynamic profile change at a certain stimulating signal. The dynamic profiles were shown in figure 22.

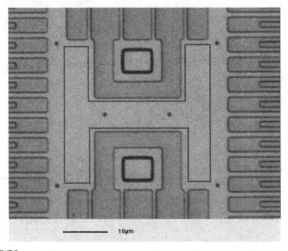

Fig. 21. Choice of ROI

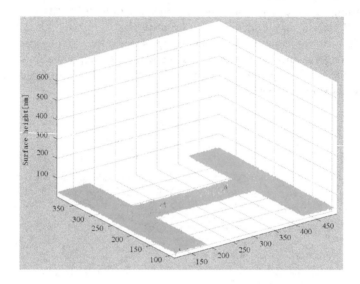

(a) Profile at the second phase

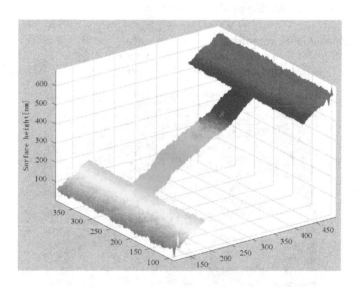

(b) Profile at the ninth phase

Fig. 22. Dynamic profile measurement

The stimulating voltage frequency was 23 kHz, the amplitude was 10 V, and the offset voltage was 20 V. The motion can be classified into three kinds of cases: 1) one reference plane; 2) the first phase of the device movement; 3) the reference mirror in the interferometer. Here the second case is chosen. From figure 22, it can be shown that the movement of the device is like sinusoidal waveform and there is a distort phenomena in the process of the movement.

5. Conclusion

This chapter has discussed several optical methods to realize MEMS characterization, including dimensional (static) and moving (dynamic) properties analysis. Two measuring systems are introduced. NMM based system combines nano-measuring machine with high positioning accuracy and microscopic interferometers. PSI, WLI, WLPSI are applied in the system to measure the dimensional parameters. MMA combines video microscopy, stroboscopic illumination and special algorithms. It employs two kinds of nondestructive methods - computer microvision for in-plane motion measurement and phase shifting interferometry for out-of-plane motion measurement. This system can test the three-dimensional motions and dynamic profiles with nanometer accuracy.

6. Acknowledgment

The authors gratefully acknowledge the support of National Natural Science Fund (91023022), International Cooperation Project of MOST (2008DFA71610) and Natural Science Fund of Tianjin (09JCYBJC05300).

7. References

Bosseboeuf, A., Gilles, J. P., Danaie, K. et al.(1999). A versatile microscopic profilometer-vibrometer for static and dynamic characterization of micromechanical devices. *Proc. SPIE*, Vol. 3825, pp. 123-133

Burdess, J. S., Harris, A. J., Wood, D. et al. (1997). A system for the dynamic characterization of microstructures. *J. Microelectromechan. Syst.*, Vol.6, No.4, pp. 322-328

Fu, X., Liu, Y. Q., Hu, X. D. et al. (2004). Micro Laser Doppler Vibrometer Technology for MEMS Dynamic Measurement. *J of Optoelectronics ·Laser*, Vol.15, No.11, pp.1357-1360

Guo, T., Chang, H., Chen, J.P. et al. (2009). Micro-motion Analyzer used for Dynamic MEMS Characterization, *Optics and Lasers in Engineering*, Vol. 47, No.3-4, pp. 512-517

Guo, T., Wu, Z. C., Ma, L. et al.(2010). Dynamic MEMS characterization system using differential phase measurement method, *Proceedings of SPIE*, Vol. 7544, pp. 75444W

Guo, T., Ma, L., Zhao, J. et al. (2011). A nanomeasuring machine based white light tilt scanning interferometer for large scale optical array structure measurement, *Optics and Lasers in Engineering*, Vol. 49, No.9-10, pp.1124-1130

Guo, T., Ma, L., Chen, J.P. et al.(2011). MEMS surface characterization based on white light phase shifting interferometry, *Optical Engineering*, Vol. 50, No.5, pp. 053606

Hartzell, A. L., Woodilla, D. J. (2001). MEMS reliability, characterization, and test. *Proc. SPIE*, Vol. 4558, pp. 1-5

Ma, L., Guo, T., Yuan, F. et al.(2009). Thick film geometric parameters measurement by white light interferometry, *Proceedings of SPIE*, Vol. 7507, pp. 75070G

6

Surface Characterization and Interfacial Adhesion in MEMS Devices

Y. F. Peng and Y. B. Guo
Xiamen University
China

1. Introduction

The characteristic size of MEMS is ranging from atomic and molecular scales to micrometer and several millimeters scales. Components that reach micro-scale size have a high surface to volume ratio, which leaves them be highly subjected to micro-scale effect and susceptible to surface forces. Devices that utilize MEMS technology will often having mating surfaces. Adhesion force can arise from any number of phenomenon such as van der Waals, capillary, ionic and molecular forces. The components used in MEMS structures are very light (on the order of a few micrograms) and operate under very light loads (on the order of a few micrograms to a few milligrams). Surface forces between the adjacent surfaces are becoming dominant over the inertial force in MEMS devices. Because of the micro-sized component, the adhesion forces can pull the adjacent compliant structure into contact and result the interfacial adhesion, which may cause the device-malfunction to a great extent. The operation and performance of lightly loaded micro/nano components in MEMS are highly dependent on the adhesive interactions between mating surfaces. In a word, it is important that the mechanisms of interfacial adhesion should be explained, and separating techniques should be added to the design of MEMS scale components to ensure there is no unwanted contact. Furthermore, the interfacial adhesion between two adjacent mating surfaces is determined by the interaction of rough surfaces. The surface is all rough though in different range. The interactions among different asperities are complicated because the surface topography is consisting of so many asperities. It stands to reason that the proper surface characterization is necessary to elucidate the interfacial adhesion.

The interfacial adhesion is the science and technology of two interacting surfaces in relative motion and of related subjects and practices. It is also valuable in the fundamental understanding of interfacial phenomena to provide a bridge between science and engineering in MEMS. The differences between the conventional or macro-contact and micro/nano-adhesion are contrasted in Table 1. In macro-contact mechanics, tests are conducted on components with relatively large mass under heavily loaded conditions. In these tests, contacting between mating surfaces is inevitable and the bulk properties of mating components dominate the contacting performance. In micro/nano-adhesion, measurements are made on components, at least one of the mating components, with relatively small mass under lightly loaded conditions. The interaction is not limited only to the contacting condition. In this situation, though without contact, the attractive interaction between mating opposite surfaces at small approaching distance can't be neglected. Some of the smaller asperities on the micro-sized surface will be stretched, while some of the taller ones will be compressed through contact. The classical contact mechanics is no longer valid in analyzing the interaction of mating micro/nano-sized surface. It is necessary to explore methods to solve the interfacial adhesion problems in MEMS devices.

Macro-contact mechanics		Micro/Nano adhesion	
Condition	Contact (Inevitable)	Condition	Contact or non-contact
	Large mass		Small mass
	Heavy load		Light or zero load
Method	Hertz theory	Method	Need to consider the adhesion of surface forces
	Linear elastic mechanics		
Target	Bulk material	Target	Surface (Few atomic layers to several μm depths)

Table 1. Comparison between macro-contact and micro/nano-adhesion

In this chapter, we will take a close look at surface geometric structure, or surface topography, and surface forces to elucidate the adhesion problems between mating micro-sized MEMS surfaces. Firstly, the complexities of the surface microstructure devices are discussed. Secondly, several typical surface-measurement instruments are introduced. Thirdly, the techniques to characterize the complex micro-scale surfaces are presented. Finally, the surface forces are described in a summary form, and then the adhesion models are given to interpret the adhesive interaction of MEMS devices.

2. Characterization and modeling of microstructure surface

2.1 Complexities of surface microstructure

Whether a surface is rough or smooth, the answer is — it depends on a roughness sensors used (Bushan, 1999)! The problem of scale-dependent roughness is very intrinsic to solid surfaces. For most solid surfaces it is observed that under repeated magnification, more and more roughness keeps appearing until the atomic scales are reached where roughness occurs in the form of atomic steps (Williams & Bartlet, 1991) as shown graphically in Fig.1. The roughness often appears random and disordered, ranges from around 10^{-4}m (0.1 mm) to about 10^{-9}m (1 nm) and does not seem to follow any particular structural pattern (Thomas, 1982). The randomness and the multiple roughness scales both contribute to the complexity of the surface geometric structure.

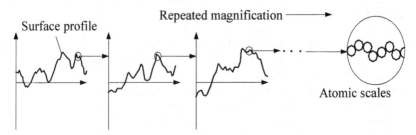

Fig. 1. Scale-dependent of surface roughness (Bhushan, 1999).

2.2 Surface measurement techniques

Because of the complexities of surface microstructure, the measuring techniques and instruments are important to achieve the surface information to characterize and model the surface microstructure in MEMS devices. The accuracy of traditional contact (probing) as well as noncontact techniques has been perfected to a level allowing measurement of roughness in the nanometer range (Fig.2). The most accurate profilometer probes allow measurement of summit heights of several Angstroms (Bennet & Dancy, 1981; Bhushan et al., 1988). Yet, the comparatively poor lateral (horizontal) resolution significantly limits application of these techniques to the nanometer topographies when the distance between asperities is much less than the solution or 0.1-1μm. The development of techniques using probes smaller than the radius of the probing needle or the light wavelength makes it possible to extend the spectrum of surfaces studied (Myshkin et al., 2003). The scanning electron microscope (SEM) technique can be used to gauge topography with a comparable resolution both vertically and laterally by interpreting the emission intensity of the secondary electrons the topographic pattern (Myshkin et al., 1992). The scanning tunneling microscope (STM) has a still finer probe, which is the electron flux tunneled between the target surface and the needle tip. In this case the surface topography resolution is 0.01 and 0.1 nm in the vertical and lateral directions, respectively (Binnig & Rohrer, 1982). Hence, the STM technique and others resulting from its progress make it possible to use this approach for more accurate topographic investigations of solids on the nanoscale. Significant prospects are connected with the application of atomic force microscope (AFM) (Sarid, 1991) in which atomic-molecular surface effects are registered.

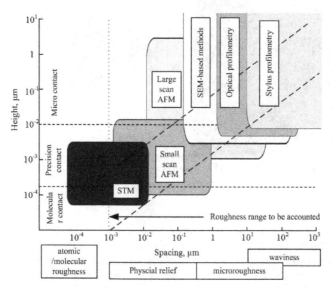

Fig. 2. Diagram of the height and spacing parameters and ranges of vertical-lateral resolution for different methods of roughness measurement (Myshkin et al., 2003).

The scanning tunneling microscope (STM) developed by Dr. Gerd Binnig and Heinrich Rohrer has revolutionized the study of surfaces and is rapidly becoming a required tool in almost every surface characterization laboratory. It is the first instrument capable of directly obtaining three-dimensional images of solid surfaces with atomic resolution (Binnig et al., 1982). Today's STMs can be used in the ambient environment for atomic-scale imaging of surfaces. Generally, samples to be imaged with STM must be conductive enough to allow a few nanoamperes of current to flow from the bias voltage source to the area to be scanned. In many cases, nonconductive samples can be coated with a thin layer of a conductive material to facilitate imaging. AFM can be used for measurement of all engineering surfaces which may be either electrically conducting or insulating. AFM has now become a main surface profiler for topographic measurements on micro- to nanoscale (Bhushan & Blackman, 1991; Oden et al., 1992; Bhushan et al, 1997). STMs, AFMs, and their modifications can be used at extreme magnifications ranging from 10^{-3} to $10^{-9} \times$ in x-, y-, and z-directions for imaging macro- to atomic dimensions with high-resolution information and for spectroscopy (Bushan, 1999). These instruments can be used in any sample environment such as ambient air (Binnig & Smith, 1986), various gases (Burnham et al., 1990), liquid (Marti et al., 1987; Binggeli et al., 1993), vacuum (Binnig et al., 1982), low temperatures (Hug et al., 1993), and high temperatures. To decrease the wear of brittle tip and extend its application in biological research, the carbon nanotube (CNT) has been used to probe the sample instead by adhered it on top of a tip (Fang et al., 2008) in AFM. The resolution ratio can reach about 3nm with functional single-walled CNT in scanning the grease double molecular membranes, while it is about 15nm for conventional Si and of Si3N4 (Yamachika et al., 2004). Fig.3 shows such a tip-CNT probe and the captured image.

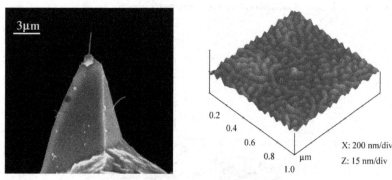

Fig. 3. AFM CNT probe and captured image (Fang et al., 2008); (a). SEM images of CNT probe (b). Images of a styrene-ethylene/butylene-styrene copolymer

2.3 Surface characterization techniques

The characterization of the surface roughness on the micro/nanoscale needs more thorough investigation. This is essential for solving interfacial adhesion phenomena. The randomness suggests that the statistical methods of roughness characterization should be adopted to determine the average dimensions of topographical elements forming the surfaces of solids. In addition, a rough surface involves so many length scales ranges from atomic/molecular level to nano or micro scale, then the characterization techniques must be independent of any length scale. In this section, both the statistical and fractal method to characterize surface roughness are presented in a way that is suitable to model adhesion of mating MEMS devices.

2.3.1 Probability height distribution

In "Handbook of Mirco/Nano Tribology", Bhushan has summarized various theories of probability distribution of rough surface (Bhushan, 1999). One of the characteristics of a rough surface is the probability distribution (Papoulis, 1965). It is often found that the normal or Gaussian distribution fits the experimentally obtained probability distribution quite well (Thomas, 1982; Bhushan, 1990). In addition, it is simple to use for mathematical calculation (Greenwood & Williamson, 1966; Chang et al., 1987). The bell-shaped normal distribution (Papoulis, 1965) which has a variance of unity is given as

$$g(\bar{z}) = \frac{1}{\sqrt{2\pi}} \exp\left[-\frac{(\bar{z}-\bar{z}_m)^2}{2}\right] \qquad -\infty < \bar{z} < \infty \tag{1}$$

where $\bar{z}_m = \bar{z}/\sigma$ is the nondimensional mean height, σ is the standard deviation. The mean height and the standard deviation can be found from a roughness measurement $z(x,y)$ as

$$z_m = \frac{1}{L_x L_y} \int_0^{L_x} \int_0^{L_y} z(x,y) dx dy = \frac{1}{N_x N_y} \sum_{i=1}^{N_x} \sum_{j=1}^{N_y} z(x_i,y_j) \tag{2}$$

$$\sigma = \sqrt{\frac{1}{L_x L_y} \int_0^{L_x} \int_0^{L_y} [z(x,y) - z_m]^2 dx dy} = \sqrt{\frac{1}{N_x N_y} \sum_{i=1}^{N_x} \sum_{j=1}^{N_y} [z(x_i,y_j) - z_m]^2} \tag{3}$$

Here, L_x and L_y are the lengths of surface sample, whereas N_x and N_y are the number of points in the x and y lateral directions, respectively. The integral formulation is for theoretical calculations, whereas the summation is used for calculating the values from finite experimental data.

Although used extensively, the normal distribution has limitations in its applicability. The normal distribution near the tail is not an accurate representation of real surfaces. This is an important point since it is usually the tail of the distribution that is significant for calculating the real area of contact (Bushan, 1999). The inverted chi-squared (ICS) distribution fit the experimental data much better near the tail of the distribution (Brown & Scholz, 1985). This is given for zero mean and in terms of nondimensional height, $\bar{z}$, as

$$g(\bar{z}) = \frac{(v/2)^{v/4}}{\Gamma(v/2)} (\bar{z}_{max} - \bar{z})^{(v/2)-1} e(\bar{z} - \bar{z}_{max}) \sqrt{v/2} \qquad -\infty < \bar{z} < \bar{z}_{max} \tag{4}$$

which has a variance of $2v$ and a maximum height $\bar{z}_{max} = \sqrt{v/2}$. The advantage of the ICS distribution is it has a finite maximum height, as does a real surface, and has a controlling parameter v, which gives a better fit to the topography data. It is found that as v increases, the ICS distribution tends toward the normal distribution (Bushan, 1999; Brown & Scholz, 1985). Berry and Hannay (Berry & Hannay, 1978) suggested that the variance can be represented as follows:

$$\sigma^2 \approx L^n \tag{5}$$

where L is the length of the sample and n varies between 0 and 2.

If the exponent n in Equation (5) is equal to zero, then the rough surface is generally to be a statically stationary process. This means that the measured roughness sample is a true statistical representation of the entire rough surface. However, n is not equal to zero in the general cases. Then a rough surface is assumed to be a nonstationary random process and the standard deviation is scale dependent, which arises from the probability distribution of a small surface region may be different from that of the larger one. The gathered roughness measurements of a wide range surfaces by Sayles and Thomas (Sayles & Thomas, 1978) have shown that the variance of the height distribution is a function of the sample length and in fact suggested that the variance varied as $\sigma^2 \approx L$. This behavior implies that the surface is a nonstationary process and any length of the surface cannot fully represent the surface in a statistical sense.

Other statistical parameters, such as rms slope σ' and rms curvature σ'' proposed by Nayak (Nayak, 1971, 1973) are also used in surface roughness characterization (Greenwood & Williamson, 1966; Nayak, 1973) and to model the elastic-plastic contact of isotropic and anistropic solid bodies (McCool, 1986). The question is that the determination of σ, σ' and σ'' depends on the sample size, instrument resolution, and experimental filter used to acquire the topography data (Yan & Komvopoulos, 1998), that is whether the rms parameters vary with the statistical sample size or the instrument resolution. Given a rough surface, an instrument with resolution τ will measure the surface height of points that are separated by a distance τ. If τ is reduced, new locations on the surface are accessed. Due to the multiple scales of roughness present, a reduction in τ makes the measured profile look different for the same surface. It is thus necessary to obtain some scale-independent techniques for roughness characterization.

2.3.2 Fractal techniques

It is found that the power spectra of engineering surfaces produced by random processes, such as cleavage, solidification, vapour deposition, and directionally unbiased machining, have been obaserved to follow inverse power laws over a wide range of length scales (Majumdar & Tien, 1990). This is an inherent property of fractal geometry illustrating its potential to represent surface features from the microscale down to the nanoscale (Yan & Komvopoulos, 1998). Fractal geometry, pioneered by Manderbrot (Mandelbrot, 1967) when studying the problem of the length of Britain coastline, can be observed in various natural phenomena, such as precipitation, turbulence, and surface topography, and is characterized by continuity, nondifferentiablity, and self-affinity. Recent works (Kardar et al., 1986; Gagnepain, 1986; Majumdar & Bhushan, 1990) have shown that the fractal geometry can be utilized to develop a scale-independent characterization technique of the fractallike behavior for a rough surface. The mathematical properties of fractal geometry can be satified by the Weierstrass-Mandelbrot (W-M) function given by (Berry & Lewis, 1980)

$$w(x) = \sum \gamma^{(D-2)n} \left(1 - e^{i\gamma^n x}\right) e^{i\phi_n} \tag{6}$$

where w is a complex function of the real variable x. A fractal profile $z(x)$ can be obtained as the real part of $w(x)$

$$z(x) = \text{Re}\left[z(x) \right]$$

$$= \sum_{n=-\infty}^{\infty} \gamma^{(D-2)n} \left[\cos\phi_n - \cos(\gamma^n + \phi_n) \right] \tag{7}$$

where D ($1 < D < 2$) is the fractal dimension of the profile, is a frequency index, ϕ_n is a random phase, and γ ($\gamma > 1$) is a parameter that determines the density of frequencies in the profile, which is often chosen to be 1.5. The right hand side of Equation is a superposition of cosine function with geometrically increasing frequencies. The random phase ϕ_n is used to prevent the surface profile. The approximate continuous power spectrum, $P(\omega)$, of the profile $z(x)$ given be Equation

$$P(\omega) = \frac{1}{\omega^{(5-2D)} \ln\gamma} \tag{8}$$

is an inverse power function of the spatial frequency, ω, and has been observed to hold for many engineering surfaces.

The two-variable function developed by Ausloos and Berman (Ausloos & Berman, 1980) can be used to model fractal surfaces exhibiting corrugations in all directions. The height function of a fractal surface can be expressed

$$z(\rho,\theta) = \left(\frac{\ln\gamma}{M} \right)^{\frac{1}{2}} \sum_{m=1}^{M} A_m \sum_{n=-\infty}^{\infty} \left(\kappa\gamma^n \right)^{(D-3)} \cdot \left\{ \cos\phi_{m,n} - \cos\left[\kappa\gamma^n \rho \cos(\theta - \alpha_m) + \phi_{m,n} \right] \right\} \tag{9}$$

where D ($2 < D < 3$) is the fractal dimension of the surface. The physical significance of D is the extent of space occupied by the rough surface, with larger D values corresponding to denser profiles. For isotropic surfaces, the value of D can be determined from the slope of the log-log plot of power spectrum (Wang & Komvopoulos, 1994; Gagnepain & Rogues-Carmes, 1986). The parameter M denotes the number of supposed ridges used to construct the surface. The anisotropy of the surface geometry is controlled by the magnitude of A_m. For isotropic surface, $A_m = A$ for all m values; for anisotropic surfaces, A_m varies with m. The arbitrary angle α_m is used to offset the ridges in the azimuthal direction. The parameter κ is a wave number related to the sample size, $\kappa = \pi m/M$. The frequency index n is a finite value. The lowest frequency of index $n_{\min}$ is equal to $1/L$ and it can also be set equal to zero. The upper limit of n is

$$n_{\max} = \text{int}\left[\frac{\log(L/L_s)}{\log\gamma} \right] \tag{10}$$

where $\text{int}[...]$ denotes the maximum integer value of the number in the brackets, L_s is the cut-off length of sample.

By introducing a new length parameter G such that G, the surface height function of 3D isotropic surfaces can be obtained

$$z(x,y) = L\left(\frac{G}{L}\right)^{D-2}\left(\frac{\ln\gamma}{M}\right)^{\frac{1}{2}}\sum_{m=1}^{M}\sum_{n=0}^{n_{max}}\gamma^{(D-3)n}\cdot\left\{\cos\phi_{m,n}-\cos\left[\frac{2\pi\gamma^{n}(x^{2}+y^{2})^{\frac{1}{2}}}{L}\cdot\cos\left(\tan^{-1}\left(\frac{y}{x}\right)-\frac{\pi m}{M}\right)+\phi_{m,n}\right]\right\}$$ (11)

The Equation (11) can be used to represent a 3D isotropic fractal surface. This function of surface height provides a deterministic means of generating stochastic rough surfaces. The only unknown variables in Equation (11) are the scale dependent fractal parameters G and D, which can be determined experimentally. Therefore, this fractal approach has the inherent capability of representing surfaces at various length scales, different from those at which the measurements were made (Yan & Komvopoulos, 1998).

3. Surface forces and adhesion mechanics

Surface microstructures typically range from 0.1 to several µm in thickness with lateral dimensions of 10-500µm, and lateral and vertical gaps to other structures or to the substrate of around 1µm (Maboudian & Howe, 1997). The large surface area and small offset from adjacent surfaces makes these microstructures especially vulnerable to adhesion upon contact. The causes of strong adhesion can be traced to the interfacial forces existing at the dimensions of microstructures. These include capillary, electrostatic, van der Waals, and chemical forces.

3.1 Surface forces and adhesion work
There are a wide variety of surface forces (Israelachvili, 1992). Capillary, electrostatic and van der Waals forces can each contribute to adhesion under different circumstances in MEMS devices.

3.1.1 van der Waals forces
Van der Waals force is the force acting between atoms or small molecules, which includes dispersion force, Debye force and dipole-dipole force. The interaction potential between atoms or molecules of each force is a function of $1/r^{6}$, which r is the separation between atoms. For two flat parallel surfaces, and for separations less than a characteristic distance, $r_{0} \approx 20$ nm (nonretarded regime), the attractive force per unit area is given by (Israelachvili, 1985)

$$F_{vdW}(r) = \frac{A}{6\pi r^{3}}$$ (12)

where A is the Hamaker constant to reflect the strength of the van der Waals interaction for two bodies in medium. However, for separations larger than r_{0}, the attraction is retarded. Taking retardation into account, it is proposed (Cheng & Cole, 1988) that the van der Waals force per unit area is represented more accurately as

$$F_{vdW}(r) = \frac{A}{6\pi r^{3}}\frac{r_{0}}{r^{3}(r+r_{0})}$$ (13)

The work of adhesion between two surfaces interacting with each other via van der Waals interaction can be obtained by integration from contact, $r_{0,vdW}$, to infinity (Maboudian & Howe, 1997)

$$W_{vdW}(r) = \frac{A}{12\pi r_{0,vdW}^2} \tag{14}$$

3.1.2 Electrostatic forces

Electrostatic forces are the forces between charged bodies. Charges are known to accumulate from the ambient and migrate across insulating surfaces on silicon chips. Early in the development of integrated-circuit (IC) technology, charge migration was the source of device instabilities. Transport of both positive and negative ionic species has been observed in the presence of lateral electrical fields (Shockley, 1964). Electrostatic attraction may also arise due to a difference in the work function of the approaching surfaces. Neglecting the internal space charge regions, the force per unit area acting between surfaces with potential difference V separated by an air gap with permittivity ε_0 is given by

$$F_{el}(r) = \frac{\varepsilon_0 V^2}{2r^2} \tag{15}$$

and the associated energy is given by

$$W_{el}(r) = \frac{\varepsilon_0 V^2}{2r} \tag{16}$$

3.1.3 Capillary force

With the presence of a thin liquid film, such as a lubricant or adsorbed water layer at the contact interface, menisci will form around the contacting and noncontacting asperities due to surface energy effects (Israelachvili, 1985). Fig.4 (a) shows the condition that the amount of liquid film volume was large enough to immerse the rough surface. When the mating surfaces are pulled apart, meniscus will formed underneath the microstructures. Then the liquid meniscus will create a pressure because of the pressure difference across the curved liquid-air interface, which is called the capillary pressure (Laplace pressure), and is given by

$$P_l = \gamma_l \left(\frac{1}{r_1} + \frac{1}{r_2} \right) \tag{17}$$

The liquid surface tension is denoted by γ_l, and the two radii of curvature of the liquid surface are termed by r_1 parallel to the surface normal of the substrate and r_2 (in the plane of the substrate) (Adamson, 1990; Israelachvili,1985). Since in micromechanical structures lateral dimensions are often much larger than the vertical spacing, $r_2 \gg r_1$, and in this case, Equation (17) simplifies to

$$P_l = \frac{\gamma_l}{d}(\cos\theta_1 + \cos\theta_2) \tag{18}$$

where θ_1, θ_2 is the upper and lower contact angle of liquid bridge, and d is the separation distance between the two surfaces, equal to $r_1(\cos\theta_1 + \cos\theta_2)$. On a hydrophilic surface ($\theta < 90°$), such as the native oxide of silicon, the meniscus shape will be concave underneath a structure shown in Fig.4 (a). This creates an attractive capillary force that may sufficiently strong to pull the compliant structures into contact.

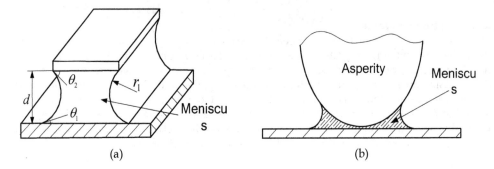

Fig. 4. Wetting and contact angle

The isolated micromenisci would occur at the contact interface, if the amount of meniscus volume were not large enough to immerse all asperities of the rough surface, as shown in Fig.4 (b), the meniscus radius at equilibrium is equal to the so-called Kelvin radius r_k. It is related to the Kelvin equation value and controlled by the relative vapor pressure relative humidity (Admoson, 1990). At equilibrium, the meniscus curvature is related to the relative vapor pressure (H_r) by the Kelvin equation:

$$\gamma_l \left(\frac{1}{r_1} + \frac{1}{r_2} \right)^{-1} \equiv r_k = \frac{\gamma_l v}{RT \log H_r} \tag{19}$$

where r_1 and r_2 are the two radii of curvature of the meniscus, r_k is the Kelvin radius, γ_l is the surface tension of the liquid, and v is its molar volume ($\gamma_l v / RT \approx 0.54$ nm for water at $20°$ C). As two hydrophilic surfaces approach each other in a humid environment, the liquid undergoes capillary condensation as soon as the separation equals

$$d_0 \approx r_k (\cos\theta_1 + \cos\theta_2) \tag{20}$$

If, after the condensation has occurred, the two surfaces are pulled apart, the volume of the condensate is essentially constant and is given by $V = S_w d_0$, where S_w is the wetted surface area (Maboudian & Howe,1997). The effect of a liquid condensate on the adhesion force per unit area between two parallel plates is then given by

$$F_{cap}(d) = \frac{\gamma_l d_0}{d^2}(\cos\theta_1 + \cos\theta_2) \tag{21}$$

If we assume that, as the two surfaces are pulled apart, the meniscus breaks at a separation much larger than d_0, then integrating Equation (21) from d_0 to infinity yields the work of adhesion due to capillary forces

$$W_{cap} = \gamma_l \left(\cos\theta_1 + \cos\theta_2 \right) \tag{22}$$

3.2 Adhesion models of single asperity

To better understand the interfacial adhesion of MEMS devices, it is important to provide an adequate background of the prior work performed in the area of adhesive rough surface contact. For determining the interfacial adhesive behavior, several solutions have been developed and many of these theories idealize the asperity in contact with a half rigid flat as a spherical shape.

Hertz theory is the famous continuum contact mode to predict the contact area for various geometries. It relates the radius of the circle of contact a_H to the load P, the spherical indenter radius R, and the equivalent elastic modulus of the contacting materials K by:

$$P_H = \frac{K a_H^3}{R} \tag{23}$$

and between the contact radius a_H and the indentation depth δ,

$$\delta = \frac{a_H^2}{R} \tag{24}$$

In the presence of surface forces, Hertz theory can underestimate the contact area, especially when the load diminishes to zero. Considering the contact between a rigid sphere with half rigid flat, the adhesion force P_a, between then is given be Bradely theory (Bradley, 1932) as

$$P_a = 2\pi\omega R \tag{25}$$

DMT theory was then proposed by Derjaguin, Muller and Toporov to account for the long-ranged attraction around the periphery of the contact area. The DMT model gives the contact radius a_{DMT} related to the work of adhesion, ω, by

$$P_{DMT} = \frac{K a_{DMT}^3}{R} - 2\pi\omega R \tag{26}$$

$$\delta = \frac{a_{DMT}^2}{R} \tag{27}$$

It is apparent that DMT is Hertz with an offset due to surface forces. Therefore, DMT theory applies to rigid systems, low adhesion and small radii of curvature. JKR theory, described by Johnson, Kendall and Roberts, takes the short-ranged attractive forces among the contact area into account. It related the contact radius, a_{JKR}, to the work of adhesion, ω, as

$$P_{JKR} = \frac{K a_{JKR}^3}{R} - \sqrt{6\pi\omega K a_{JKR}^3} \tag{28}$$

$$\delta = \frac{a_{JKR}^2}{R} - \frac{2}{3}\sqrt{\frac{6\pi\omega a_{JKR}^3}{K}} \tag{29}$$

JKR theory applies well to highly adhesive systems that have large radii of curvature and low stiffness. To bridge the DMT theory and JKR theory, by following the analysis of Tabor (Tabor, 1977; Muller et al., 1980) pointed out that the two theories represented the opposite extremes of a dimensionless parameter μ given as

$$\mu = \left(\frac{R\omega^2}{E'^2\varepsilon^2}\right) \tag{30}$$

where ε is the equilibrium spacing in the lennard-Jones potential. μ can be interpreted as the ratio of elastic deformation resulting from adhesion to the effective range of surface forces.

A more complex, yet more accurate, description of sphere–flat adhesion mechanic, which is referred as MD model, was formulated by Maugis (Maugis, 1992). By analogy with the plastic zone ahead of a crack, the adhesion is represented by a constant additive traction acting over an annular region around the contact area. The ratio of the width of the annular region to the radius of the contact area is denoted by m. The set of equations relating the dimensionless load, approach is

$$1 = \frac{\lambda A^2}{2}\left[\sqrt{m^2-1}+(m^2-2)tg^{-1}\sqrt{m^2-1}\right]+\frac{4\lambda^2 A}{3}\left[1-m+\sqrt{m^2-1}tg^{-1}\sqrt{m^2-1}\right] \tag{31}$$

$$\overline{P} = A^3 - \lambda A^2\left(\sqrt{m^2-1}+m^2 tg^{-1}\sqrt{m^2-1}\right) \tag{32}$$

$$\Delta = A^2 - \frac{4\lambda A}{3}\sqrt{m^2-1} \tag{33}$$

where λ is another dimensionless number, called transition parameter λ, and is related to μ by $\lambda = 1.157\mu$. The dimensionless parameters that appear in the above equations are defined as follows: $\lambda = \dfrac{2\sigma_0}{\left(\pi\omega K^2/R\right)^{1/3}}$, $\overline{P} = \dfrac{P}{\pi\omega R}$, $A = \dfrac{a_{MD}}{\left(\pi\omega R^2/K\right)^{1/3}}$, $\Delta = \dfrac{\delta}{\left(\pi\omega^2 R/K^2\right)^{1/3}}$, where the adhesion work ω is defined as $\sigma_0 h_0$, σ_0 is the adhesive attraction equals Dugdale stress and h_0 is the effective range of Dugdale stress.

For each previous mentioned theories were presented, there may be cases when assumptions made for a given approach do not exactly describe the materials combinations or the geometry, which are depicted physically in Fig.5.

Following the analysis of Maguis, Kim et al (Kim et al., 1998) offers an extension of the MD solution by adding to the solution regime when the asperity might not be in physical contact but still in the range of adhesion. The KMJ extensions explained by Kim is

$$1 = \frac{\pi}{4}\lambda c^2 + \frac{2}{3}c(\pi-2)\lambda^2 + \xi \tag{34}$$

$$\overline{P} = -\frac{\pi \lambda c^2}{2} \tag{35}$$

$$\Delta = -\frac{4C\lambda}{3} - \frac{2}{\pi}\frac{\xi}{\lambda} \tag{36}$$

where $C = \dfrac{c}{\left(\pi \omega R^2/K\right)^{1/3}}$, c represents the adhesive contact zone radius, ξ is ratio of h_g/h_0, and h_g, h_0 are the gap between the deformed asperity at $r=0$ and $r=c$ respectively shown in Fig.5(e.)

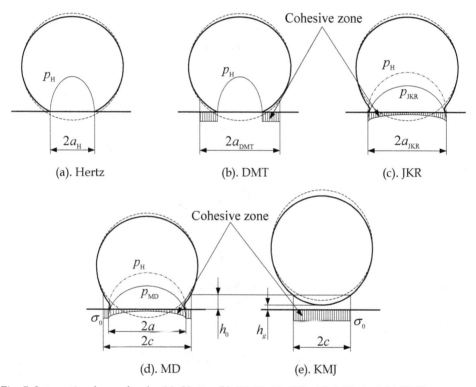

Fig. 5. Interactive forces for the (a). Hertz, (b). DMT, (c). JKR, (d). MD, and (e).KMJ.

3.3 Adhesion for micro-sized rough surface

In practice, the contact at the surface interface is governed by asperity interaction. Since the surfaces are not smooth, contact of two multiscale rough surfaces will occur only at discrete points which sustain the total compressive force. The typical contact interface which is formed of contact spots of different sizes that are spatially distributed randomly over the interface. The size of contact spots ranges from nanometers to micrometers, making adhesion a multiscale phenomenon.

3.3.1 Statistical adhesion theories for micro-sized rough surface

Most contact theories of the rough surface thus far are mostly based upon the conventional statistical parameters such as standard deviation of asperity heights, slope and radius of curvature (Greenwood & Williamson, 1966; Fuller & Tabor, 1975). The Greenwood-Williamson (GW) model assumes the surface to be composed of hemispherical asperities all having the same radius of curvature R. The summit heights or asperity peaks are distributed randomly about a mean summit plane and follow a Gaussian distribution with a standard deviation, σ. If there exists a probability density function $\varphi(z)$ of asperity heights, then it is possible to find the probability that an asperity will be greater than a certain height, d. The distance d represents the length from the mean plane of asperity heights to the smooth surface. The probability that an asperity height is greater than d is given by:

$$\int_d^\infty \varphi(z)\mathrm{d}z \tag{37}$$

Therefore, it follows that the number of asperities in contact is represented by

$$n = N\int_d^\infty \varphi(z)\mathrm{d}z \tag{38}$$

where N represents the total number of asperities.

Having the numerical expressions for the non-dimensional contact radius (A_i) and load ($\overline{P}_i$) for a single asperity as a function of Δ, the total contact area and load can be formed (Morrow et al., 2003)

$$A_{total} = N\left(\frac{\pi\omega R^2}{K}\right)^{2/3}\int_d^\infty \pi A_i(\Delta)^2\,\varphi(z)\mathrm{d}z \tag{39}$$

$$P_{total} = N\pi\omega R\int_d^\infty \overline{P}_i(\Delta)^2\,\varphi(z)\mathrm{d}z \tag{40}$$

The asperities that have a height than d greater than are deformed by a distance $\delta = z - d$. Assuming a Gaussian distribution and to have a relationship between Δ and z, the following equation can be obtained:

$$\frac{P_{total}}{N\pi\omega RP_c(\lambda)} = \int_d^\infty \overline{P}_i\left(\frac{\delta}{\overline{\delta}}\right)^2\exp\left(-\frac{(\delta+d)^2}{2\sigma^2}\right)\mathrm{d}\delta \tag{41}$$

where $\overline{\delta}$ is $\left(\pi^2\omega^2R/K^2\right)^{1/3}$. The above equation is only valid when the smooth surface progressively approaches the rough surface until a minimum d is reached [Fuller and Tabor]. Because of the existence of $\delta_c\left(-\left(3\pi^2\omega^2R/4K^2\right)^{1/3}\right)$ to abrupt rupture, or pull-off, the asperities will no longer contribute to the adherence force when asperities were extended above δ_c. Therefore, Morrow et al made the adjustment of the lower integration limit by the amount $\delta_c(\lambda)$, the adhesion model given above then takes the form:

$$\frac{P_{total}}{N\pi\omega R P_c(\lambda)} = \frac{1}{P_c(\lambda)\sqrt{2\pi}} \int_{-2\delta_c^*(\lambda)}^{\infty} \overline{P_i}\left(\frac{\delta}{\overline{\delta}}\right)^2 \exp\left(-\frac{\left(\delta^* + d^*\right)^2}{2}\right) d\delta^* \qquad (42)$$

$$\frac{A_{total}}{N\left(\pi\omega R^2/K\right)^{2/3} A_c(\lambda)} = \frac{\pi}{A_c(\lambda)\sqrt{2\pi}} \int_{-2\delta_c^*(\lambda)}^{\infty} A\left(\frac{\delta^*}{\overline{\delta}^*}\right)^2 \exp\left(-\frac{\left(\delta^* + d^*\right)^2}{2}\right) d\delta^* \qquad (43)$$

Any term in above equation that has a superscript has been divided by σ. The improvement equation by Morrow is similar in form to the rough surface integral of Fuller and Tabor, but has some important difference. The most important is that the lower integration limit δ_c and P_c are functions of λ, which gives the solution validity over the entire range of the transition parameter. P_c represents the force at which the system becomes stable under force control. The normalization factor $P_c(\lambda)$ is determined by finding the point at which the tangent for load deflection curve becomes zero. $\delta_c(\lambda)$ can also be determined in a similar manner. The critical step in obtaining an adhesive rough surface solution is to find the load at which the system becomes unstable, i. e. the minimum pull-off force (P_{min}). The d^* can be solved by set the derivative of Eq. (42) equals to zero (Morrow, 2003).

3.3.2 Fractal adhesion theories for micro-sized rough surface

The statistically based adhesive theory can be used with confidence as long as the length scale is known before hand (Morrow, 2003). It is well documented that surfaces exhibit roughness on many different length scales (Majumdar & Bhushan, 1990; Majumdar, 1989; Majumdar & Bhushan, 1991). The topography of any surface can be thought of as roughness surperimposed on top of roughness. Majumdar et al (Majumdar & Bhushan, 1990; Majumdar & Tien, 1990) have proven that the multi-scale nature for surface roughness can be represented by fractal geometry. Then it is reasonable to establish the adhesion model of rough interface based on the fractal parameters. Majumdar (Majumdar & Bhushan, 1991) has argued that the size distribution of contact spots can be given as:

$$n(s) = \frac{D}{2s}\left(\frac{s_l}{s}\right)^{D/2} \qquad (44)$$

where s is the contact area and s_l represents the largest spot contact area.

3.3.2.1 Fractal model for adhesive contact of JKR type

By assuming that the plasticity plays a minor role in the asperity contact due to the light loading conditions, Morrow et al proposed a fractal model for adhesive contact of JKR type (Morrow & Lovell, 2003). The model follows the example set forth by Majumdar and Bhushan (Majumdar & Bhushan, 1991). It was assumed that the interference which a spherical has with a rigid plane is given by:

$$\delta = G^{(D-1)}l^{(2-D)} \qquad (45)$$

where l is the length scale of the asperity as shown in Fig. 6.

A relationship between the truncated and real contact area is developed by equaling the interference distance for both the Hertizan truncated and JKR conditions:

$$s_{JKR} = 9.25 \left(\frac{s'^D G^{2(1-D)} \omega}{K} \right)^{2/3} \left(1.68 - 0.61 \sqrt{5.95 - \frac{\sqrt{(s'\pi)/2}}{\left(\left(s'^D G^{2(1-D)} \omega \right) / K \right)^{1/3}}} \right) \tag{46}$$

Following the work of Majumdar and Bhushan, the expression for the interference, δ, in terms of truncated area s':

$$\delta = G^{(D-1)} s'^{(2-D)/2} \tag{47}$$

The radius of curvature at the asperity tip in terms of fractal parameters (Majumdar & Bhusha, 1991)

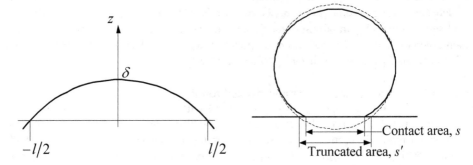

(a) Geometry of contact spot with a given interference   (b) Truncated area and contact area

Fig. 6. Fractal approximation of asperity contact (Morrow, 2003)

$$R = \frac{s'^{D/2}}{\pi^2 G^{D-1}} \tag{48}$$

Because of the surface attraction, the asperities are stretched when the contacting surfaces are pulled away. The critical interference δ_c is given

$$\delta_c = -\left(\frac{3\omega^2 s'^{D/2}}{4K^2 G^{D-1}} \right)^{1/3} \tag{49}$$

The critical contact area s'_c for adhesion is broken is:

$$s'_c = \left(\frac{3\omega^2}{4K^2} \right)^{1/(3-2D)} G^{4(1-D)/(3-2D)} \tag{50}$$

Supposing that the size distribution of the asperities, $n(s)$, is known, the real area of contact can be integrated

$$s'_{real} = \int_{s'_c}^{s'_l} S'n(s)ds' = \frac{D}{2-D}\left(s'_l - s'^{1-D/2}_c s'^{D/2}_l\right) \tag{51}$$

where s'_l represents the truncated area of the largest contact spot and is related with the total truncated area S' with

$$s'_l = \left(\frac{3-D}{D-1}\right)S' \tag{52}$$

Then the pull-off force is

$$P = \int_{s'_c}^{s'_l} P(s')n(s)ds' \tag{53}$$

The above Equation can be numerically integrated and solved (Morrow, 2003).

3.3.2.2 Fractal elastic-adhesive model

Based on the work of Yan and Komvopolous (Yan & Komvopoulos, 1998), Morrow proposed a 3D fractal elastic adhesive rough surface solution methodology (Morrow, 2003). In order to develop easy to use expressions for the asperity interference, Yan and Komvopoulos developed a two-dimensional form of Equation (11) to account for the M number of ridges by introducing a multiplicative factor which was eventually set equal to 1. This implied that a two-dimensional W-M function could be used to approximate a fractal function in three dimensions. Yan states in (Yan & Komvopoulos, 1998) that since the radius of curvature of each asperity is much greater than the height of the asperity then this relationship can be assumed to be

$$a' = 2R\delta \tag{54}$$

where a' is the truncated contact radius. Then the radius of curvature is:

$$R = \frac{s'^{(D-1)/2}}{2^{5-D}\pi^{(D-1)/2}G^{D-2}\left(\ln\gamma\right)^{1/2}} \tag{55}$$

where the truncated contact area s' can be expressed as

$$s' = \left(\frac{\delta}{2^{4-D}\pi^{(D-3)/2}G^{D-2}\left(\ln\gamma\right)^{1/2}}\right)^{2/(3-D)} \tag{56}$$

The fractal relationship for the radius of curvature, R can be used to modify the transition parameter λ which is related to fractal dimensions

$$\lambda = 1.16\left(\frac{\omega s'^{(D-1)/2}}{2^{5-D}K^2\pi^{(D-1)/2}G^{D-2}\sqrt{\ln\eta z_0^3}}\right) \tag{57}$$

where z_0 is the intermolecular distance. From the fractal adhesion model, it can be found that transition parameter λ is no more a constant, but a variant with the approaching distance δ of the asperity during the adhesive contact process. Then the truncated area for each ith asperity can be determined based on a given interference δ_i:

$$\frac{s'_i}{\pi} = 2\left(\frac{s'^{(D-1)/2}}{2^{5-D}\pi^{(D-1)/2}G^{D-2}\left(\ln\gamma\right)^{1/2}}\right)\delta_i \tag{58}$$

Equation 58 can be solved using a fixed point iteration scheme to determine the truncated area. It should be noted that the truncated area is used for the first iteration of the algorithm only. To solve for the total load and contact area, Morrow proposed a novel numerical algorithm. In this algorithm the Maugis-Dugdale solution is used to model the micro-sized contact of each asperity. The first step is to generate the surface topography from the fractal parameters using W-M function. Next the surface is offset to introduce an initial penetration into the rigid plane. The interference, δ_i, of each asperity is determined. Based on this interference, the truncated area is subsequently computed for only the asperities that are physically interfering with the rigid plane and is then used to compute the radius of curvature and transition parameter λ for each asperity. Once the truncated area is initially determined, the main iteration scheme is started to determine the real area of contact for each asperity. Using the values for λ and R, the contact radius, m and δ_{com} are computed. These values are computed based on the adhesive contact solution of Maugis given in Equations (31-33). Once the iterations have converged, the values of the load, P_i, and area, S_i are added to the totals (P_{total} and S_{total}). All interfering asperities are iterated on in this fashion and then the surface is moved to the next separation locatio1n and the procedure starts once again.

3.3.2.3 Fractal elastic-plastic adhesion model

According to Majumdar-Bhushan model (Majumdar & Bhushan, 1991), the truncated area $s' > s'_c$ are elastically deformed since they satisfy the condition of $\delta < \delta_c$, whereas asperities with $s' \le s'_c$ satisfy the plastic flow criterion and are thus considered to be in fully plastic deformation state. This result is in disagreement with that derived from the GW model. The reason for this disagreement is that the present analysis accounts for the dependence of the curvature radius on microcontact area, whereas in the GW model the curvature radius of asperity is considered to be invariant. The critical microcontact area for plastic flow of the entire asperity is

$$s'_c = \left(\frac{2^{11-2D}}{9\pi^{4-D}}G^{2D-4}\ln\gamma\left(\frac{E}{H}\right)^2\right)^{1/(D-2)} \tag{59}$$

The elastic adhesion of asperities can be analyzed by Morrow method (Morrow, 2003). Assuming the asperity is at a fully plastic microcontact, the contact pressure within the contact zone is the hardness, H. Then with the Maugis-Dugdale approximation to the adhesive interaction, the adhesive contact pressure for the microcontact can be approximated as (Peng & Guo, 2007)

$$p_p = \begin{cases} H - \dfrac{2\sigma_0}{\pi}\tan^{-1}\sqrt{\dfrac{c^2-a^2}{a^2-x^2}} & |x| \ll a \\ -\sigma_0 & a < |x| < c \end{cases} \tag{60}$$

The adherence force can be obtained by integration over the contact and cohesive zones

$$P_p = Hs - 2\sigma_0 a^2 \left[\sqrt{\frac{c^2}{a^2} - 1} + \frac{c^2}{a^2} \tan^{-1} \sqrt{\frac{c^2}{a^2} - 1} \right] \tag{61}$$

where $s = 2\pi a^2$ is the real contact area. When plastically deformed, the contact area s of the microcontact is just the truncated area, which means $a = a'$ and is

$$a = \frac{1}{2} \left(\frac{\delta}{2G^{D-2}(\ln\gamma)^{1/2}} \right)^{1/(3-D)} \tag{62}$$

The radius of cohesive zone c can also be determined by geometrical consideration

$$c = \frac{1}{2} \left(\frac{\delta + h_0}{2G^{D-2}(\ln\gamma)^{1/2}} \right)^{1/(3-D)} \tag{63}$$

If $s_l' > s_c'$, both elastic and fully plastic microcontacts exists. Thus the total adherence force for the fractal surface includes the elastic adhesion and plastic adhesion forces, which is

$$F_t = \sum_j P_{ej} + \sum_j P_{pj} \tag{64}$$

where P_{ej} is the adherence force of j th asperity in elastic contact which can be determined by MD theory utilizing the Morrow method, and P_{pj} is the adherence force of j th asperity in plastic contact which can be calculated by Equation (61).

3.3.2.4 Adhesion model for meniscus stiction

The issue of meniscus force is of critical importance for the microsized interfacial interaction mechanism, such as in the magnetic storage hard-disk drives (Bhushan, 1996). For a given interacting system and environmental parameters, the Laplace pressure and the meniscus height can be assumed to be a constant, whose character is similar to that of the Dugdale stress in linear elastic fracture mechanics. To solve the adhesive problem of the capillary force due to meniscus, the effective work of adhesion can be defined by the product of the meniscus height and Laplace pressure (Xue & Polycarpou,200; Peng et al., 2009)

$$\omega_l = h_c p_l \tag{65}$$

Then by substituting the σ_0 with p_l, and h_0 with h_c, the above methodology, mentioned in section 3.3.2.2 and 3.3.2.3, can be adopted to solve the sitiction problem in presence of meniscus.

4. Summary

In this chapter, we have attempted to present a description of issues and techniques in the interfacial adhesion for the MEMS devices. We firstly discus the complexity of the surface microstructure. Then we present the techniques to characterize the micro-scale surfaces. Finally, we introduce the adhesion models to interpret the adhesive interaction of MEMS devices.

It is hoped that the introductions in this chapter can gain the rational understanding leading to the design of better MEMS structures in the technologically field. The interpretation of interfacial adhesion is challenging for the application of MEMS technology. It is further complicated by the inability to observe the interfacial interactions directly, resulting in conclusions from inference. The gap between theoretical research of rough surface adhesion and the real world where thousands or millions of asperities are involved remains enormous. Then it is clear that there is still a great deal of research necessary to obtain a comprehensive understanding of adhesion at the microscale. The high-resolution instrument should be developed and well calibrated, with which one can measure both the microstructure topography and adhesion, especially the biological sample and hydrophilic surface. It is essential that the proper data processing method should be presented to reflect the intrinsic characters more accurately, and helps to understand the sources of error.

5. References

Adamson, A. W. (1990). *Physical Chemistry of Surfaces*, 5th ed., Wiley, New York.

Ausloos and Berman. (1980). A multivariate Weierstrass-Mandelbrot function, *Proceedings of the Royal Society A* 370, 459.

Bennet, J.M. & Dancy, J.H. (1981). Stylus profiling instrument for measuring statistical properties of smooth optical surfaces, *Applied Optics* Vol. 20: 1785–1802.

Berry, M. V. & Hannay, J. H. (1978). Topography of Random Surfaces, *Nature* Vol. 271: 573.

Berry, M. V. & Lewis, Z. V. (1980). On the Weierstrass-Mandelbrot Fractal Function, *Proceedings of the Royal Society A* Vol. 370: 459-484.

Binggeli, M., Christoph, R., Hintermann, H.E., Colchero, J., Marti, O. (1993). Friction Force Measurements on Potential Controlled Graphite in an Electrolytic Environment, *Nanotechnology* Vol. 4: 59–63.

Binnig, G., Quate, C. F. and Gerber, C. (1986). Atomic Force Microscope, *Physics Review Letter* Vol. 56: 930–933.

Binnig, G. & Rohrer, H. (1982). Scanning tunneling microscopy, *Helvetica Physica Acta* Vol. 55: 726-735.

Binnig, G. & Smith, D.P.E. (1986). Single-Tube Three-Dimensional Scanner for Scanning Tunneling Microscopy, *Review of Scientific Instruments* Vol. 57: 1688.

Bhushan, B. (1999). *Handbook of Micro-Nanotribology*, Second Edition. Bharat Bhushan. CRC Press.

Bhushan, B. (1996). *Tribology and Mechanics of Magnetic Storage Devices*, 2nd ed., Springer, New York.

Bhushan, B. & Blackman, G.S. (1991). Atomic Force Microscopy of Magnetic Rigid Disks and Sliders and Its Applications to Tribology, *Journal of Tribology* Vol. 113: 452–458.

Bhushan, B. & Dugger, M. T. (1990). Real Contact Area Measurements on Magnetic Rigid Disks, *Wear* Vol. 137: 41–50.

Bhushan, B., Sundararajan, S., Scott, W.W., and Chilamakuri, S. (1997). Stiction Analysis of Magnetic Tapes, *IEEE Transactions on Magnetics* Vol. 33: 3211–3213.

Bhushan, B., Wyant, J.C., Meiling, J. (1988). A new three-dimensional non-contact digital optical profiler, *Wear* Vol. 122: 301–312.

Bradley, R. S. (1932). The coercive force between solid surfaces and the surface energy of solids, *Philosophical Magazine* Vol. 13: 853-862.

Brown, S. R. and Scholz, C. H. (1985). Closure of Random Elastic Surfaces in Contact, *Journal of Geophysical Research* Vol. 90: 5531-5545.

Burnham, N.A., Domiguez, D.D., Mowery, R.L., Colton, R.J. (1990). Probing the Surface Forces of Monolayer Films with an Atomic Force Microscope, *Physics Review Letter* Vol. 64: 1931-1934.

Chang, W. R., Etsion, I., and Bogy, D. B. (1987). An Elastic-Plastic Model for the Contact of Rough Surfaces, *Journal of Tribology* Vol. 109: 257-263.

Cheng, E. & Cole, M. W. (1988). Retardation and Many Body Effects in Multilayer Film Adsorption, *Physics Review B* Vol. 38: 987-995.

Fang, F. Z., Xu, Z. W. and Dong, S. (2008). Study on phase images of a carbon nanotube probe in atomic force microscopy, *Measurement Science & Technology* Vol. 19(No. 5) doi:10.1088/0957-0233/19/5/055501.

Fuller, K. & Tabor, D. (1975). The effect of surface roughness on the adhesion of elastic solids, *Proceedings of the Royal Society A* Vol. 345: 327-342.

Gangepain﹐ J. & Roques-Carmes. (1986). Fractal Approach to Two-Dimensional and Three-Dimensional Surface Roughness, Wear Vol. 109: 119-126.

Greenwood, J. A. & Williamson, J. B. P. (1966). Contact of Nominally Flat Surfaces, *Proceedings of the Royal Society A* Vol. 295: 300-319.

Hug, H.J., Moser, A., Jung, Th., Fritz, O., Wadas, A., Parashikor, I., Guntherodt, H.J. (1993). Low Temperature Magnetic Force Microscopy, *Review of Scientific Instruments* Vol. 64: 2920-2925.

Israelachvili, J. N. (1985). *Intermolecular and Surface Forces*. Academic, London.

Kardar, M., Parisi, G., and Zhang, Y. C. (1986). Dynamic Scaling of Growing Interfaces, *Physics Review Letter* Vol. 56: 889-892.

Kim, K.S., McMeeking, R.M. and Johnson. K.L. (1998). Adhesion, slip, cohesive zones and energy fluxes for elastic spheres in contact, *Journal of the Mechanics and Physics of Solids* Vol. 46:243-266.

Maguis, D. (1992). Adhesion of spheres: The JKR-DMT transition using a Dugdale model, *Journal of Collid and Interface Science* Vol. 150(No. 1): 243-269.

Maboudian, R. & Howe. R. T. (1997). Critical Review: Adhesion in surface micromechanical structures, *Journal of Vacuum Science & Technology B* Vol. 15(No.1): 1-20.

Majumdar, A. (1989). *Fractal surfaces and their Applications to Surface Phenomena*. PhD thesis, University of California, Berkley.

Majumdar, A. & Bhushan, B. (1991). Fractal model of elastic-plastic contact between rough surfaces, *Journal of Tribology* Vol. 113:1-11.

Majumdar, A. and Bhushan, B. (1990). Role of Fractal Geometry in Roughness Characterization and Contact Mechanics of Surfaces, *Journal of Tribology* Vol. 112: 205-216.

Majumdar, A. & Tien, C. L. (1990). Fractal Characterization and Simulation of Rough Surfaces, *Wear* Vol. 136: 313-327.

Mandelbrot, B. B. (1967). How Long is the Coast of Britain? Statistical Self-Similarity and Fractional Dimension. *Science* Vol. 155: 636-638.

Marti, O., Drake, B., Hansma, P.K. (1987). Atomic Force Microscopy of Liquid-Covered Surfaces: Atomic Resolution Images, *Applied Physics Letter* Vol. 51: 484-486.

McCool, J. I. (1986). Comparison of Models for the Contact of Rough Surfaces, *Wear* Vol. 107: 37–60.

Morrow, A. (2003). *Adhesive Rough Surface Contact*, Ph.D. thesis, University of Pittsburgh, Pittsburgh.

Morrow, C., Lovell, M.R., Ning, X. (2003). A JKR–DMT transition solution for adhesive rough surface contact, *Journal of Physics D Applied Physics* Vol. 36: 534–540.

Myshkin, N.K., Ya. A., Grigoriev, S.A., Chizhik, Choi, K.Y., Petrokovets, M.I. (2003). Surface roughness and texture analysis in microscale, *Wear* Vol. 254: 1001–1009.

Myshkin, N.K., Ya., Grigoriev, A., Kholodilov, O.V. (1992). Quantitative analysis of surface topography using scanning electron microscopy, *Wear* Vol.153 (No.1): 119–133.

Muller, V. M., Yushchenko, V. S., B. V. Derjaguin. (1980). On the influence of molecular forces on the deformation of anelastic sphere and its sticking to a rigid plane, *Journal of Colloid Science on Science* Vol. 77: 91-101.

Nayak, P. R. (1971). Random Process Model of Rough Surfaces, *Journal of Lubrication Technology* Vol. 93: 398–407.

Nayak, P. R. (1973). Random Process Model of Rough Surfaces in Plastic Contact, *Wear* Vol. 26: 305–333.

Oden, P.I., Majumdar, A., Bhushan, B., Padmanabhan, A., and Graham, J.J. (1992). AFM Imaging, Roughness Analysis and Contact Mechanics of Magnetic Tape and Head Surfaces, *Journal of Tribology* Vol. 114: 666–674.

Papoulis, A. (1965). *Probability, Random Variables and Stochastic Processes*, McGraw Hill, New York.

Peng, Y. F. & Guo, Y. B. (2007). An Adhesion Model for Elastic-Plastic Fractal Surfaces, *Journal of Applied Physics* Vol. 102(No.5): 3510-7.

Peng, Y. F., Guo, Y. B., Hong, Y. Q. (2009). An Adhesion Model for Elastic-contacting Fractal Surfaces in Presence of Meniscus, *ASME Journal of Tribology* Vol. 131: 024504-1-5.

Sarid, D. (1191). *Scanning Force Microscopy*, Oxford University Press, New York.

Sayles, R. S. and Thomas, T. R. (1978). Surface Topography as a Nonstationary Random Process, *Nature* Vol. 271: 431–434.

Shockley, W., Hooper, W.W., Queisser, H.J. and Schroen, W. (1964). Mobile electric charges on insulating oxides with application to oxide covered silicon p-n junctions, *Surface Science* Vol. 2: 277-287.

Tabor, D. (1977). Surface forces and surface interactions, *Journal of Colloid Science on Science* Vol. 58: 2-13.

Thomas, T. R. (1982). *Rough Surfaces*, Longman, New York.

Wang, S. & Komvopoulos, K. (1994). A fractal theory of the interfacial temperature distribution in the slow sliding regime: Part I – Elastic contact and heat transfer analysis, *Journal of Tribology* Vol. 116: 812 - 823.

Williams, E. D. & Bartlet, N. C. (1991). Thermodynamics of Surface Morphology, *Science* Vol. 251: 393-400.

Xue, X. & Polycarpou A. (2007). An improved meniscus surface model for contacting rough surfaces, *Journal of Colloid and Interface Science* Vol. 311: 203–211.

Yamachika, R. et al. (2004). Controlled Atomic Doping of a Single C60 Molecule, *Science* Vol. 304:281-284.

Yan, W. & Komvopoulos, K. (1998). Contact analysis of elastic-plastic fractal surfaces, *Journal of Applied Physics* Vol. 84: 3617.

7

Advanced Surfactant-Modified Wet Anisotropic Etching

Bin Tang and Kazuo Sato
Nagoya University
Japan

1. Introduction

In the area of Micro Electro Mechanical Systems (MEMS), bulk micromachining and surface micromachining are two main technologies. Bulk micromachining defines structures by selectively etching inside a substrate while surface micromachining uses a succession of thin film deposition and selective etching on top of a substrate. The two technologies are quite different, resulting in different dimensions and different mechanical properties. Although bulk micromachining is usually considered to be the older technology, the two developments run parallel. This is due to the fact that the two different approaches have trade-offs. Taking an example of MEMS capacitance accelerometers, surface micromachined structures use smaller chip area, thus leaving more space for the electronics. On the other hand, in bulk micromachining the larger mass gives greater sensitivity for accelerometers and the larger area leads to larger capacitances for easy read out, which are extremely useful in the inertial device fabrication.

1.1 Wet anisotropic etching in MEMS

Bulk micromachining technology relies on isotropic or anisotropic liquid phase (wet) etching as well as by plasma phase (dry) etching in single crystalline silicon for the formation of functional shapes and patterns in many major applications. Although dry etching has penetrated the traditional territory of wet etchants, the high cost in dry etching and the difficulty in the etch rate uniformity on the whole wafer in wet isotropic etching still make wet anisotropic etching the most affordable method for the reliable production, if wet anisotropic etching can be used to deliver a similar intermediate or final structure.

The formation of crystal facets due to etching is referred to as faceting. When the primary flat of Si {100} is along the [110] direction, rectangular structures with concave corners are easily made with four (111) sidewalls and a (100) plane as the bottom. If the slow etching (111) planes meet, etching will be self-limiting to result in inverted pyramids. Etched grooves, trenches, wells and other basic structures of diaphragms (membranes), beams, and cantilevers exemplify the features of crystal plane-dependent etching. Combined with the use of mask patterns, the etch rate anisotropy becomes a most valuable property as it provides a low-cost, precise means for the production of three-dimensional shapes delimited by smooth, shiny facets, leading to complex structures with multiple functionalities, as shown in Fig. 1.

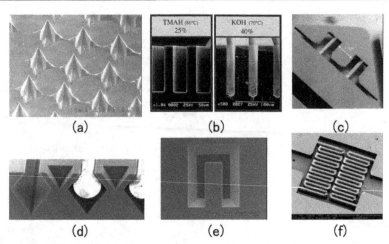

Fig. 1. Wet anisotropic etched structures for various kinds of applications: (a) microneedles for bio-medical applications, Shikida et al. 2006; (b) deep grooves, Sato et al. 1998; (c) mass-spring systems for accelerometers, Butefisch et al. 2000; (d) grooves for the optical fibre alignment, Hoffmann et al. 2002; (e) cantilever beam; (f) suspended filament beams as MEMS heaters, Lee et al. 2009.

A number of alkaline etchants have been tried for wet anisotropic etching of single crystal silicon. Some main features of wet etchants are compared in Table 1. EDP (also referred to as EPW for ethylene diamine, pyrocathecol and water) is not wildly used because of occupational safety and health hazards. Therefore, potassium hydroxide (KOH) and tetra methyl ammonium hydroxide (TMAH) are the most commonly used anisotropic etchants. Based on the ability to withstand the chemical attack by these etchants, silicon oxide (SiO_2), silicon nitride (Si_3N_4), and other metal layers (e.g. Cr, Au) have been used as masking materials. KOH is non-toxic, easy to use, provides excellent etching profiles and has a good selectivity between Si and Si_3N_4, although poor for SiO_2. In addition, KOH is incompatible with CMOS processing due to the presence of an alkali metal. Although TMAH has a lower etch rate, it has outstanding characteristics, such as a high selectivity between Si and SiO_2, and the absence of harmful ions. Hence, TMAH solutions are preferred in recent research and production.

Etchant	KOH (40 wt%)	TMAH (25 wt%)	EDP (80 wt%)
Rate (at 80 °C) μm/min	1	0.5	1 (at 115 °C)
Etching of SiO_2 mask	0	++	+
Compatibility for IC process	-	+	+
Handling (toxicity)	+	+	-
Cost	++	+	+

+:Good 0:Fair -:Poor

Table 1. Comparison of some main features of wet etchants.

1.2 Motivation of surfactant-modified etchants

With different etchants or in the conditions of different concentrations and temperatures, people could get different etching characteristics, which can be clarified as etch rate anisotropy, surface roughness and mask-corner undercut. Surface roughness improvement is important when considering the optical and tribological applications. On the other hand, the conventional design of MEMS structures fabricated by wet silicon bulk micromachining on Si {100} has sharp edge convex and concave corners. This design exhibits stress concentration at the concave corners when a load is applied, which may initiate micro cracks. By providing rounded concave corners instead of sharp ones, the stress can be reduced, thus improving the mechanical efficiency of the microstructures. In pure TMAH solutions, however, both surface roughness and mask-corner undercut are not good enough in regard to above applications. Fig. 2 illustrates anisotropic etched cantilever beam shaped patterns with mask-corners in 10 wt% TMAH at 60 °C. The etched surface is full of hillocks with an average roughness of 110nm. Therefore, many other factors have to be thought over.

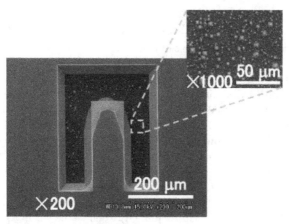

Fig. 2. Anisotropic etched cantilever beam shaped patterns with mask-corners in 10 wt% TMAH at 60 °C (etching depth = 34 μm). The upper right figure is 100 μm x 100 μm.

Among those methods for improving surface roughness and mask-corner undercut, for instance, metal impurities, alcohols, diffusion, light, pressure, microwave irradiation, corner compensation and ultrasonic irradiation et al., surfactant-modified wet anisotropic etching is more outstanding for the effects in both surface smoothness and undercut decrease, also for its stabilization of anisotropy change.

Various ionic (e.g. anionic SDSS, cationic ASPEG, etc.) and non-ionic (e.g. PEG, NC series, Triton X-100, etc.) surfactants have been investigated. Although anionic surfactants exhibit the highest etching rate, the cationic and non-ionic surfactants are suitable for TMAH solutions to improve the roughness of the etched surface owing to the excellent capacity to wet the silicon wafer. TMAH solutions with cationic and non-ionic surfactants are IC-compatible process. Furthermore, adding non-ionic surfactant to TMAH solutions can efficiently reduce undercutting at mask-corners. Such an addition is preferred when accurate profiles are required without very deep etching. Therefore, non-ionic surfactant-modified etching process attracts researchers' attention. With regard to easily handling and less toxicity, in this study, Triton X-100 is selected.

This chapter starts by providing the completely etch rate anisotropy in surfactant-modified wet etching in section 2; the mechanism behind of the change of etching characters when compared with pure etchants will be analyzed in section 3; several applications for the fabrication of new structures by using this advanced anisotropic wet etching will be presented in section 4.

2. Characterization of the etch rate anisotropy

Different etchants can give different etching properties. As one of the most important properties, the etch rate anisotropy clearly manifests the etching behavior in a concentrated solution. In this chapter, the characterization of the etch rate anisotropy is studied by using hemispherical silicon specimens, with and without surfactant in TMAH solutions. Especially, surfactant-modified etching process is analyzed in detail because of its benefit in MEMS applications.

2.1 Experimental details

The hemispherical specimen enables us to obtain the etch rate for all range of crystallographic orientations under the same etching conditions simultaneously, because all the orientations are placed on its surface. The P-type single-crystal hemispherical specimen with a diameter of 44 mm (resistivity: 6–12 Ω-cm) is used in the evaluation. The ingot of hemispheres is provided by Sumitomo Sitix Corporation. In order to produce hemispheres, it is mechanically ground, lapped and polished into mirrored surfaces with a sphericity of less than 10 μm, latitude from 0 to 90° and surface roughness of 0.005–0.007 μm in the arithmetical average by Okamoto Kogakukosakusho Corporation. The etch rate at each orientation is calculated by measuring the shape change before and after etching. The optimized etching depth should be in the range of 100–150 μm in order to avoid interference between neighboring orientations while maintaining the resolution of the geometry measurement. The shape is measured using a 3D profile machine UPMC550-CARAT (Carl Zeiss Co.) with an accuracy of less than 1.0 μm. Fig. 3 shows the locations of crystallographic orientations on the silicon hemispherical sample and a schematic view of the surface profile measurement. Area 'A' corresponds to the measurable area of the hemispherical silicon sample. The place outside the measurement zone and the bottom area are protected by a thermally grown oxide layer. The surface profile is probed every 2° of latitude ranging from 0° to 70°, and every 2° of longitude ranging from 0° to 360°. Supplements of de-ionized (DI) water into an etching bath every 2 h control the tolerance of etching temperature within 1 °C. The total numbers of probe points to be measured are 6480.

TMAH (Toyo Gosei Co. Ltd) and Triton X-100 (Amersham Biosciences) are used as the main etchant and surfactant, respectively. The Triton solution is used to prepare the surfactant-added TMAH solution. The fresh etchant is employed in every subsequent experiment.

2.2 Etch rate anisotropy

The photos of the hemispherical specimen before and after etching at 61 °C in TMAH + Triton are taken as one of the examples and presented in Fig. 4. The contour maps made by the 3D Anisotropic-Etching Simulator (FabMeister-ES) are shown in Fig. 5. Due to the crystallographic symmetry of the hemispherical specimen, only one quarter part with averaged etch rates among equivalent orientations is presented. The range of the planes affected by surfactant adding is clearly visible. The etch rates of exact and vicinal {100} planes are almost unaffected when the surfactant is added, while the etch rates of exact and vicinal {110} planes are reduced significantly.

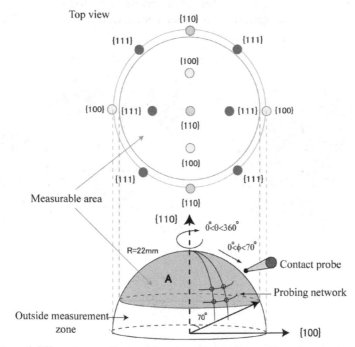

Fig. 3. Location of different crystallographic orientations on one silicon hemispherical sample and schematic view of surface profile measurement. (Reproduced with permission from IOP)

Fig. 4. Photos of the hemisphere specimen: (a) before etching and (b) after etching (etching temperature = 61 °C).

The effect of etching temperature on the etch rates is shown in Fig. 6. The etching anisotropy is influenced by temperature. The orientation of the highest etch rate is shifted toward {350} with the increase in temperature. Contrary to 71 and 81 °C, no single plane prominently appears as the highest etching rate plane at 61 °C. The etch rates of the orientations between {100} and {110} largely depend on the temperature; however, those in between {110} and {111} exhibit less dependence. It can obviously be concluded from these results that the etching anisotropy depends upon the etching temperature. This will result in different etched profiles due to the difference in etching temperatures even if the same mask is used. Examples of measured etch rates for nine different orientations are listed in Table 2. The etch rate ratios of {mnl}/{100}, where m, n, and l are integers, increase with temperature for the planes lying between {100} and {110} (excluding {100} and {110}). On the other hand, these ratios for the planes lying between {110} and {111} are almost the same for 61 °C; however, for other temperatures (71 and 81 °C), the values vary.

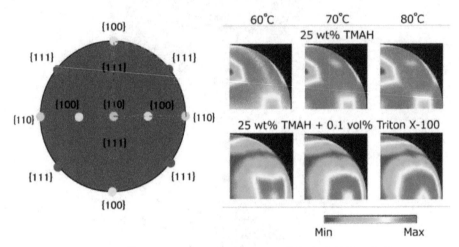

Fig. 5. Contour maps of etch rate in pure and surfactant-added TMAH solutions

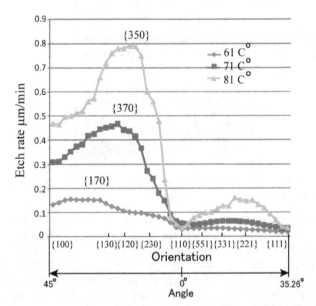

Fig. 6. Effect of temperature on the etch rate anisotropy in 25 wt % TMAH + 0.1 vol% Triton X-100.

Orientation	61°C		71°C		81°C	
	Etch rate (μm/min)	{mnl}/{100}	Etch rate (μm/min)	{mnl}/{100}	Etch rate (μm/min)	{mnl}/{100}
{100}	0.133	1.000	0.308	1.000	0.468	1.000
{130}	0.138	1.038	0.462	1.500	0.717	1.532
{120}	0.102	0.767	0.437	1.419	0.709	1.515
{230}	0.093	0.699	0.272	0.883	0.600	1.282
{110}	0.032	0.241	0.055	0.179	0.036	0.075
{551}	0.034	0.256	0.055	0.179	0.098	0.209
{331}	0.034	0.256	0.064	0.208	0.126	0.269
{221}	0.033	0.248	0.062	0.201	0.152	0.325
{111}	0.019	0.143	0.033	0.107	0.035	0.077

Table 2. Orientation- and temperature-dependent etch rates and etch rate ratios of {mnl}/{100} in TMAH + Triton solutions.

Temperature-dependent etching anisotropy is verified by fabricating 3D circular shape cavities on silicon {100} wafers. Fig. 7 shows the SEM images of circular cavities (etch depth = 100 μm) formed in Triton-added 25 wt% TMAH at 61 and 81 °C using circular shape mask opening, respectively. SEM images are taken after removal of the oxide masking layer. In the case of TMAH + Triton, the {110} and its vicinal planes emerge along <100> directions as these planes exhibit significantly low etch rates. The cavity etched at 81 °C provides better roundness than that of etched at 61 °C. This is because of improved anisotropy ({mnl}/{100}) at high temperature (Table 2). When the temperature is increased from 61 to 81 °C, the etch rate ratios of {110}/{100} and {111}/{100} are changed from 0.241 and 0.143 to 0.077 and 0.075, respectively.

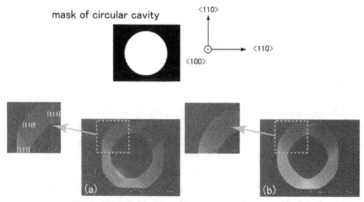

Fig. 7. Circle-like 3D microstructures in Si {100} wafers using one-step etching in 25 wt % TMAH + 0.1 vol% Triton X-100 at: (a) 61 °C and (b) 81 °C.

The Arrhenius-type dependence of etching rates for six planes {110}, {100}, {111}, {331}, {711}, and {221} is plotted in Fig. 8. The activation energy of TMAH + Triton (0.1-0.2 eV) is lower than that of pure TMAH (or KOH) solution (0.5-0.7 eV). The extreme resemblance of activation energy for {331} and {221}, which lie between {110} and {111}, may be associated with the almost same etch rates as described earlier. In particular, {110} plane has very low activation energy.

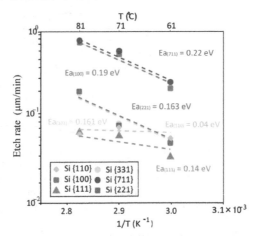

Fig. 8. The Arrhenius plot of etching rates for different planes.

3. Adsorption of surfactant molecules and its effect on etching

The change in the etching behavior of TMAH after the incorporation of the surfactant indicates that the reaction mechanism at the surface is affected by the surfactant molecules, which may block selectively some of the active surface sites, particularly those that appear on Si {110}. Triton X-100 has the following molecular structure:

$$C_8H_{17} \longrightarrow \bigcirc \longrightarrow O(CH_2CH_2O)_nH$$

hydrophobic chain   hydrophilic chain

(head)   (tail)

One end of the molecule is hydrophilic, and the other is hydrophobic.

3.1 Adsorption of surfactant molecules on silicon surfaces

Ellipsometry is utilized as a surface-sensitive technique in order to determine the thickness of thin films adsorbed on different surfaces at the subnanometer. Although the non-ionic surfactants do not take part in the etching reaction, their molecules are known to adsorb on the silicon surface by FT-IR observation. In this section, we report detailed ellipsometry measurements in order to determine and characterize the preferential adsorption of the non-ionic surfactant Triton X-100 on {110} and {100} surfaces before etching. The study focuses on the dependence of the thickness of the surfactant layer on various conditions such as the surfactant bath time (SBT), its temperature (T) and the use of ultrasonic agitation (UA), etc.

For experiments, we use three-inch diameter, p-type single crystalline {100} and {110} silicon wafers of 5-10 Ω-cm resistivity. These orientations are selected because of two reasons: (i) most widely used and commonly available, (ii) surfactant effects are quite different for Si {100} and Si {110}. Firstly, the wafers are diced into 13×13 mm² small samples. Thereafter, the samples are properly cleaned in chemical solutions followed by thorough rinse in de-ionized (DI) water. In this study all the containers are either glass or Teflon, depending on the type of solution. The samples are now stored in DI water. A surfactant bath consisting of 1 vol% Triton X-100 in DI water is prepared.

The density of adsorbed surfactant molecules in the layer increases as the surfactant concentration is increased, reaching a maximum value (adsorption saturation) at a concentration that is similar in magnitude to the critical micelle concentration (CMC), typically between 0.01 vol% and 0.1 vol% (i.e. 100–1000 ppm). Increasing the surfactant concentration further results in a larger number of micelles in solution but does not affect the adsorption density at the interface. The Triton concentration value of 1 vol% in DI water is a simple choice in order to study the effect of the pre-adsorbed layer as it ensures that enough surfactant is adsorbed on the surface according to the CMC argument.

The samples are dipped in the surfactant bath at room temperature and 60 °C for different times. Prior to immersion in the Triton bath, the samples are dipped in 5% hydrofluoric (HF) acid solution for 1 min and then thoroughly rinsed in DI water. This step is attempted to make the surface hydrophobic. After the surfactant bath, the samples are gently dipped in DI water for several times to remove the most weakly adsorbed surfactant molecules from the surface. These dippings are carried out in still water in a Teflon container. The number of dippings in DI water after the surfactant bath may affect the thickness of the adsorbed layer. Therefore, the effect of the number of gentle dippings is also studied. Moreover, the effect of ultrasonic agitation during the surfactant bath is also investigated. After several dippings in DI water, the samples are dried and used for surfactant layer thickness measurements by ellipsometry. Bare silicon samples are used as reference surfaces. For the ellipsometry measurements of the obtained Triton films we used a standard geometry where the sample is placed horizontally and visible light is reflected at a grazing angle and received by the detector. The spectral analysis is performed using a commercial spectroscopic ellipsometry analyzer (MARY-102).

Although the thickness of the Triton layer (h) increases with the Surfactant Bath Time (SBT) and the thickness can be reduced by performing one or several dippings (ND), we have observed that two Triton layers of equal thickness that have been prepared in different ways, namely, (i) by only the Triton bath and, (ii) by the Triton bath followed by several dippings, have different deviations for the measured thickness in the ellipsometry measurements. When the thickness is determined by focusing the light beam on different regions of a sample without rinsing, the measurements exhibit large fluctuations (±6 Å), indicating an uneven surfactant distribution (large salience). However, the scattering range of the thickness for the samples that have been dipped is less than ±2 Å, thus indicating a more homogeneous packing of the surfactant molecules. For improved repeatability and control, we conclude already at this stage that the Triton layer should be prepared by following the second procedure (Triton bath followed by several dippings).

In order to evaluate how the thickness of the Triton layer decreases with the number of dippings, we consider a Triton layer that is initially saturated, meaning that the silicon surface has been exposed to the Triton solution for a sufficiently long time (24 h), thus ensuring that the thickness has reached its maximum value. Fig. 9(a) shows the thickness of surfactant layer as a function of the number of dippings. Here, ND = 0 means that the sample is directly dried by air. The plot shows that, for both {100} and {110} oriented wafers, a layer of finite, non-zero thickness is measured, indicating that the surfactants are adsorbed on the silicon surface. Generally, {110} samples show a thicker Triton layer, indicating a larger ability to attract surfactant molecules. Below ND = 3, the difference between Si {110} and Si {100} at the same number of dippings is small, and the thickness is high for both orientations. For these particular measurements, the thickness changes significantly between ND = 2 and ND = 3, although it does not change much between ND = 1 and ND = 2. Above ND = 3, the thickness remains finite and constant for {110}, suggesting that the surfactant layers are strongly adsorbed –at least they cannot be easily removed by rinsing– while the thickness is vanishingly small for {100}, indicating a weaker adsorption as the surfactant layer can be easily removed. Thus, we conclude that ND = 3 is the optimal number of dippings and use it for all subsequent experiments. For ND = 3, the typical deviation in the layer thickness is roughly ±1 Å.

The surfactant layer thickness as a function of the Surfactant Bath Time (SBT) is presented in Fig. 9(b) (only ND = 3 is concerned) for two different surfactant bath temperatures (Room Temperature and 60 °C). The adsorption of the surfactant on both Si {100} and Si {110} are saturation processes, characterized by a saturation time (τ_{sat}) and a saturation thickness (h_{sat}). For non-ionic surfactants, the adsorption kinetics involve an initial fast depletion of the surfactant molecules immediately bordering the interface, followed by the diffusion of surfactant molecules from the bulk etchant to the interface and, finally, a rearrangement of the adsorbed surfactant molecules into a final packing structure. Although the actual shape of the monotonic increase of the thickness before saturation should contain valuable information about the manner how the surfactant is packed into a layer, such a study is out of the scope of the present work. From the figure, it is apparent that for the nearly same temperature {100} develops a thinner surfactant layer than {110}. The saturation thickness will become larger for smaller ND values. This observation is good in conformity with the less temperature-dependence of non-ionic surfactant adsorption.

Fig. 9(c) shows an Arrhenius plot of the saturation thickness against inverse temperature. The Arrhenius equation ($h \propto e^{-Ea/KT}$) gives the dependence of adsorption on absolute temperature and activation energy. We find that the apparent activation energy is Ea ~ 0.15 eV for {110} while Ea ~ 0.01 eV for {100}, showing that there is clearly bigger barrier to be overcome in the adsorption process of surfactant molecules on the {110} surface. The larger activation energy for {110} suggests that chemisorption may play a role on these surfaces, whereas the small activation energy for {100} indicates that only physisorption is involved for these surfaces.

From the Arrhenius plot, it is observed that adsorption is dominantly thermal for {110}, while nearly athermal for {100}. The increase in adsorption density with temperature is mostly due to the reduction in the size of the hydration shells surrounding the surfactant molecules, especially around the hydrophilic polyethylene oxide chains. The adsorption process is a trade-off between (i) the energy reduction obtained through adsorption by increasing the number of contacts between the hydrophobic alkyl chains and the surface, and (ii) the entropy increase obtained by remaining dissolved in a more disordered state. The hydration shells are smaller at higher temperatures, resulting in less water becoming ordered per adsorbed molecule, thus allowing a larger number of adsorbed surfactant molecules at higher temperature.

Agitation is a key method that can significantly affect the wet etching quality, including the etch rate and surface morphology. The etching properties of etched surfaces with ultrasonic agitation are satisfactory and superior to no agitation. Fig. 9(d) shows the surfactant layer thickness obtained using 110 W ultrasonic cleaner (VS-D100) during residence in the surfactant bath at room temperature. After adding ultrasonic agitation, the surfactant layer thickness for {110} increases with respect to no agitation (cf. figure 3.4), indicating that appropriate forced convection can improve the adsorption of surfactant molecules.

However, for {100} the thickness is only slightly larger than without agitation, indicating little change of the surface properties. The error bars are included in figure 3.6 to stress the larger variations as compared to no agitation.

Although our study shows that a larger adsorption of Triton molecules can be obtained by using ultrasonic agitation, especially on {110}, the oscillating force results in large fluctuations in the surfactant thickness. An inhomogeneously adsorbed layer is considered a disadvantage in terms of repeatability and surface roughness control.

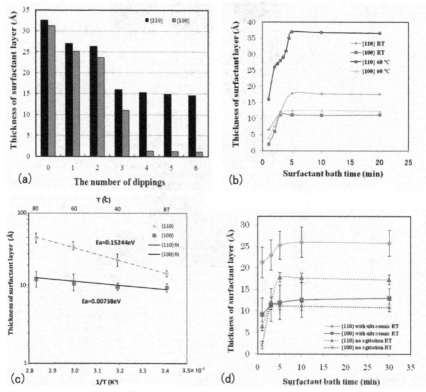

Fig. 9. Ellipsometry study: (a) thickness of the surfactant layer obtained in 1 vol% Triton as a function of the number of dipping in DI water at Room temperature; (b) surfactant layer thickness as a function of Surfactant Bath Time (SBT); (c) Arrhenius plot of the saturation thickness of surfactant; (d) comparison of the thickness of surfactant attached on the silicon surface with ultrasonic agitation at room temperature (Reproduced with permission from Elsevier).

3.2 Effect of the adsorbed surfactant layer on etching

In order to observe the effect of the adsorbed surfactant layer on the etched silicon surface morphology and etch rate, 10 wt% TMAH etchant is used at two different temperatures (room temperature and 60 °C, for both the Triton bath and the etching in TMAH) with ND = 3. The etch rates and etched surface morphology of {110} and {100} with pre-adsorbed Triton layer of different thicknesses are shown in Fig. 10 (etch depth ~ 30 ± 3 μm).

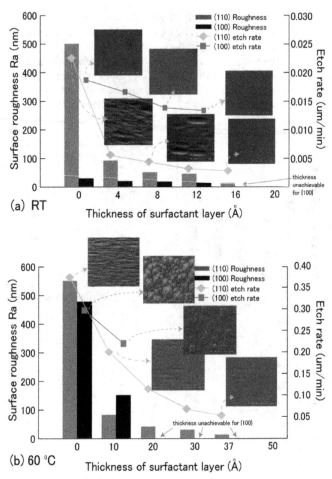

Fig. 10. Effect of pre-adsorbed surfactant layer (1 vol% Triton in DI water) on the surface roughness (Ra) and etch rate for Si {110} and Si {100} in 10 wt% TMAH: (a) room temperature; and (b) 60 °C.

For room temperature, the dramatic transformation in the surface morphology correlates directly with a strong reduction in the measured surface roughness. For {110}, it is found that typical zigzag structures emerge in pure TMAH while short pre-treatment in Triton (producing a layer thickness of 12 Å) drastically smoothens the surface, in spite of the reduced thickness of the surfactant layer. The saturated layer thickness for {110} (≈ 16 Å) produces a very smooth silicon surface (Ra ≈ 20 nm). For {100}, the surfactant pre-treatment provides some improvement in the morphology, even when the initial surface is already very smooth. Similar experiments carried out at 60 °C are shown in Fig. 10(b) (etch depth ~ 35 ± 3 μm). Although {110} shows similar surface morphologies at 60 °C and RT, the surface morphology of {100} is very different, characterized by the formation of pyramidal hillocks. Nevertheless, the roughness of both {110} and {100} is improved by the use of the surfactant layer.

In a similar manner, the thickness of the surfactant layer above the silicon surface has an effect on the etch rate in TMAH solutions after the surfactant pretreatment. For {110}, there is a significant reduction in the etch rate for thin surfactant layers. However, for {100} the reduction in the etch rate is only moderate. While {110} has a higher etch rate than {100} in pure TMAH in Fig. 10(b), the pre-treated samples show an opposite behavior, with {100} faster than {110}. This behavior is similar as for directly etching in solutions of Triton added TMAH, indicating that the dissolved surfactant is adsorbed on the surface during etching, as recently shown in FT-IR experiment. Compared with room temperature, there is a less sudden reduction in the etch rate of Si {110} wafers at 60 °C. In the case of {100}, a moderate etch rate reduction is observed at 60 °C (as for room temperature). The saturated layer thickness for {110} ($\approx$ 37 Å) produces a very smooth silicon surface (Ra $\approx$ 20 nm or less).

4. Applications on MEMS

Significantly different etching behaviour of TMAH + Triton from that of traditionally used anisotropic etchants is very useful for MEMS applications in order to extend the range of 3D structures fabricated by wet etching because the surfactant is adsorbed at the silicon-etchant interface as a thin layer to act as a filter moderating the etching behaviour. In this chapter, we present three applications using surfactant-modified etchants and point out its great potential on advanced MEMS structures.

4.1 Conformal structures

The corner compensation method is the most widely used method for fabricating the sharp edge convex corners. The design and dimensions of the compensation structures require the knowledge of the undercutting ratio and its dependence on the etchant. If the design of MEMS structures does not include any rounded concave and/or sharp convex corners but a smooth etched surface is necessarily required, then the high concentration (20 - 25 wt% TMAH) should be selected for anisotropic etching. If the structures comprise rounded concave and sharp edge convex corners, the pure TMAH cannot be used due to severe undercutting.

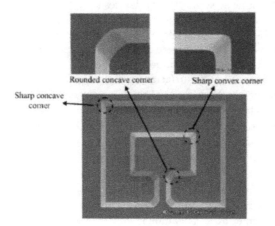

Fig. 11. Conformal mesa shape fabricated in surfactant-modified solutions.

The sharp convex corners and rounded concave corners etched in surfactant-modified TMAH solutions are shown in Fig. 11. Various kinds of etched patterns (etch depth about 35 μ m) using 25 wt% TMAH without and with the surfactant are shown in Fig. 12. The undercutting at the convex corners is considerably reduced because of the changed etching anisotropy by the adsorption of surfactant molecules. This solution exhibits minimum undercutting and provides very smooth surfaces while keeping a reasonable etch rate.

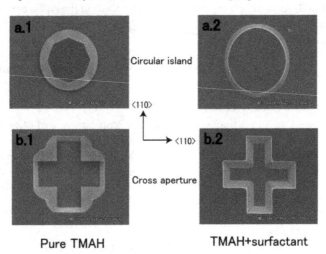

Fig. 12. Two kinds of etched patterns in pure and surfactant added 25 wt% TMAH at 60 °C (etch depth = 35 μm).

4.2 Scalloping removal for vertical micro-lens

Recently, the optical elements perpendicular to the substrate, for instance the cylindrical lens for micro-laser-scanning module applications, have been required as they can be integrated to a desired position on the same substrate using standard semiconductor processes. In order to fabricate some class of devices such as a micro-lens for optical MEMS, a highly smooth rounded column surface is required. Basically commonly used wet etchants (e.g. pure KOH and TMAH) provide very high etch rates at a rounded profile and exhibit a quite rough surface as the many faceted orientations appear on such kinds of surfaces. Hence, the removal of scalloping on rounded column surfaces is a challenging issue.

Due to the selective adsorption of surfactant molecules, the etch rate of {110} in TMAH + Triton solutions is significantly reduced. The etch rates of the planes lying between {111} and {110} at 61°C are almost same with values less than 0.05 μm. This property has been studied to remove the scalloping at the sidewalls of DRIE etched patterns on {110} silicon wafers. The study of scalloping removal is performed by short time dipping of DRIE etched patterns in TMAH + Triton. In order to investigate the effect of crystallographic directions on the surface roughness and the final shape of etched profile, a ring shape structure with internal and external diameters 200 μm and 300 μm respectively is fabricated by DRIE of about 40 μm deep. Thereafter, wet anisotropic etching is employed for 10 min in 25 wt% TMAH + 0.1 vol% Triton at 61°C. This time is long enough to clearly observe the change in profile. The SEM pictures of resultant profile and the surface morphologies of its sidewalls at different locations are shown in Fig. 13. The qualitative analysis (SEM image) reveals that the portion between two {111} planes centered by <110> orientation (marked by F), where corrugation patterned is disappeared, is sufficiently smooth for optical applications. This is because that the planes appearing in this portion exhibit very slow and almost same etch rates for 61 °C. Due to symmetry, the portion Q also has the smooth surface. This property is very useful for the fabrication of cylindrical lens, which can be easily integrated with other optical elements on a silicon substrate.

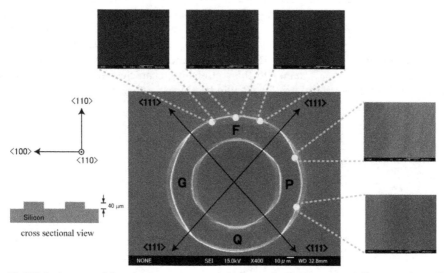

Fig. 13. SEM pictures of the ring structure and the morphologies on its different side walls after wet anisotropic etching at 61 °C.

The details of cylindrical lens portion are schematically illustrated in Fig. 14(a). The total range of the portion within two {111} planes on {110} surface is 70.54°. Since the {111} planes are the slowest etch rate planes in wet anisotropic etchants, in order to maintain the roundness of the curved profile, these planes should not be included in the design. Hence, the curved portion making an angle of 70°, is only useful portion for achieving smooth rounded vertical walls using DRIE followed by TMAH + Triton treatment. In this work, in order to demonstrate the application of surfactant-added TMAH for the removal micro-corrugation at rounded surface and the fabrication of cylindrical lens with smooth surface, the curved portion with hatched lines is selected.

The etching time for scalloping removal should be controlled accurately in order to achieve highly smooth etched surface finish while maintaining desired shape profile. In this work, the mechanism of scalloping removal is that the top area of the corrugation is etched with highest etch rate. Firstly, the left equation corresponds to the calculation of wet etching time in order to remove micro-corrugation at {110} side wall (Fig. 14(b) A-A'), where T is the etching time, $R_{\{170\}}$ is the highest etching rate at 61 °C, α is the angle between {110} and {170}, and H_1 is the height of corrugation. At 61 °C, the etching rates for {170} and {110} are the largest and smallest respectively. {170} project the corrugation, and the projected region is etched with the highest etching rate. On the other hand, {110} appears at the bottom of the corrugation, minimizing the over-etching of the sidewall. The projected region of the corrugation is becomes small as the etching process, and it finally flattens out. In our case, α is measured as $37°$ and H_1 is 69.428 nm. The etching time of 42 seconds is obtained. Secondly, removing of micro-corrugation on other planes {ijk} is shown in Fig. 14(b) B-B'. Here micro-corrugation caused by DRIE is considered as the same scalloping shape. Therefore, the angle β between {ijk} and {xyz}, where {xyz} has the highest etch rate between {110} and {ijk}, is $\beta = \alpha = 37°$. Also, the height $H_2 = H_1 = 69.428$ nm. $R_{\{ijk\}}$ is the etching rate of those planes located between {110} and {111}. Now the etching formula is described as the right equation. As noted before, etching rate of planes lying between {110} and {111} at 61 °C exhibit little variation. For this reason, we consider $R_{\{ijk\}} = R_{\{110\}}$. Another important parameter is the etch rate of highest etch rate planes in corrugated structures. Luckily, the distributions of etching rates on the {100} vicinal planes exhibit nearly identical property, which implies $R_{\{xyz\}} = R_{\{170\}}$. Thanks to the wet etching characterization in this intriguing etchants. It is feasible for the fabrication of cylindrical lens.

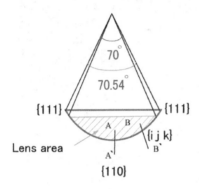

(a) Top view of cylindrical lens

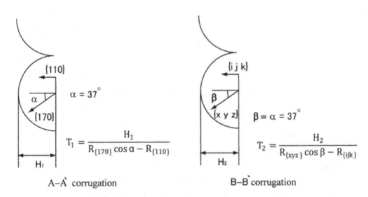

(b) Mechanism of corrugation removing

Fig. 14. (a) Schematic top view of cylindrical lens and (b) Mechanism of micro-corrugation removal by wet anisotropic etching at the sidewalls of cylindrical lens fabricated by DRIE.

Fig. 15 shows the SEM pictures and AFM measurement results of silicon surface roughness before and after wet etching. The roughness of sidewalls has been improved to about 1 nm.

This surface roughness is sufficiently enough for optical MEMS (MOEMS) applications. Thus, the small time etching in 25 wt% TMAH + 0.1 vol% Triton at 61 °C successfully remove the micro-corrugation without altering the desired shape of the DRIE etched profile and provides almost homogeneous surface finish.

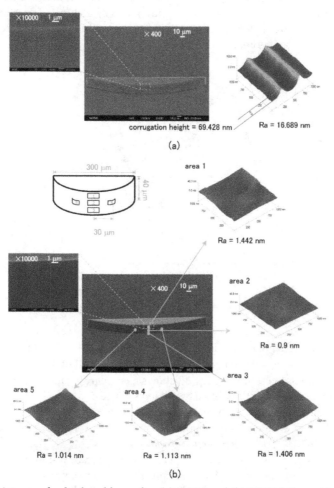

Fig. 15. SEM pictures of cylindrical lens after (a) DRIE and (b) DRIE followed by short time etching treatment in 25 wt% TMAH + 0.1 vol% Triton at 61 °C. The AFM results reveal surface roughness at different locations of vertical sidewalls. Dimension of each zoomed figure is 9 μm x 9 μm (Reproduced with permission from IOP).

4.3 Sharp tips with high aspect ratio

(111) planes are often employed as self-stop sides in the development of MEMS structures. However, on (111) silicon wafers, if certain designated planes are exposed by dry etching, novel structures dominated by fast planes might be formed in the subsequent wet anisotropic etching. In the etchant of TMAH + Triton X-100, although (110) and its vicinal

planes are strongly affected by the adsorption of surfactant molecules, having lower etch rates than other planes, there is still a local maximum plane (221) located between (110) and (111) planes at 80 °C, as shown in Fig. 16. That means (221) planes are capable of being applied to tips as the undercutting sides. The data are obtained from hemispherical samples. As we mentioned previously, there is some difference of etch rates on stressed and flat surfaces, but influence from hemispheres is limited and the relative value among the planes is sufficiently small to be ignored.

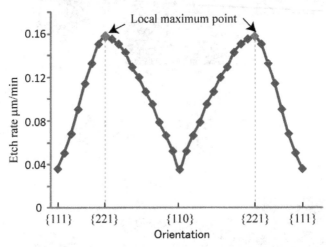

Fig. 16. Etch rate distribution between (111) and (111) in 25 wt% TMAH + 0.1 vol% Triton X-100 at 80 °C.

Here, we use a mask of equilateral triangle on (111) silicon with each side length of 60 μ min the formation of silicon tips. Wafers are firstly deep dry etched of 50 μ m and then dipped into TMAH + Triton X-100 solutions. Fig. 17 shows a circularly graphic net of (111)-centered crystal orientations analyzing the wet etched shapes. Once dry etched, six fast etching (221) planes become dominant until they meet together. The angle from periphery to (221) is about 10°, leading to a much more oblique slope than conventional ones surrounded by (311) or (411) on (100) silicon wafers etched in pure KOH or TMAH. Therefore, it is allowed to fabricate silicon tips with very high aspect ratio in surfactant-modified etchants. The picture of etched sample, not over-etched, permits the examination of etched planes. The top view of one etched tip after 18 min in the TMAH + Triton X-100 at 80 °C, as shown in Fig. 18, exhibits that the etched corner is mainly composed of two planes which have fast and same etch rates. The included angle between two adjacent lines is measured as 150±2°. Note that it is on (111) silicon and the vertex angles are directed at <112>. Making the crystalline projection as a reference, this angle indicates those orientations are located at the red line. Moreover, as mentioned above, planes with the local maximum etch rates would control the control the final. Here, in the solution of TMAH + Triton, this point lies in between (111) and (110), i.e., in blue line. The junction near (221) means those planes constituting the structure which also are in the same crystal class. This result is in good persistence with previous design, indicating the truth of a new tip with very high aspect ratio.

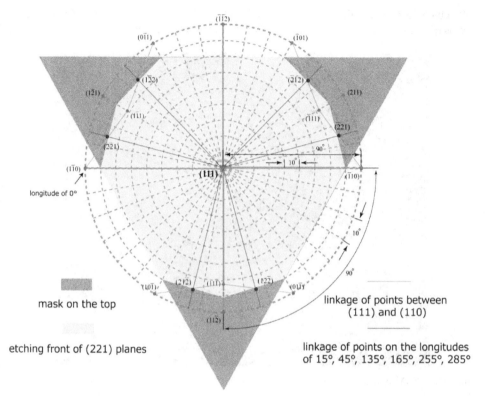

mask on the top

etching front of (221) planes

linkage of points between (111) and (110)

linkage of points on the longitudes of 15°, 45°, 135°, 165°, 255°, 285°

Fig. 17. A circularly graphic net of (111)-centered crystal orientations, illustrating the etched plane of (221).

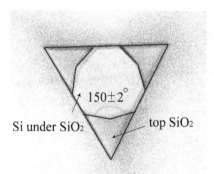

Fig. 18. One photo from optical-microscope reflecting the intermediate status during wet etching with a triangular mask on Si (111) in 25 wt% TMAH + 0.1 vol% Triton at 80 °C.

Fig. 19 shows an SEM image of the etched tip after 33 min, on which there is still a mask cap left. Furthermore, the completed tip is exhibited in Fig. 20. It is clear that the tip has a height of 30 μm, leading to the high aspect ratio of 6:1, and the front angle is measured as about

18°, which conforms with the included angle of two (221) planes. Moreover, the curvature radius of less than 10 nm is achieved without another oxidation-sharpening treatment.

Fig. 19. An SEM image of the etched tip with a cap.

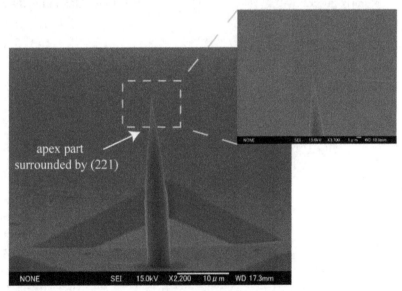

Fig. 20. A completely etched sharp tip of high aspect ratio.

5. Conclusion

In this chapter, firstly the etch rate anisotropy in surfactant-modified etch solution is investigated, showing intriguing properties that are different from that of pure alkaline solutions. The etch rates of exact and vicinal {100} planes are almost unaffected when the surfactant is added, while the etch rates of exact and vicinal {110} planes are reduced significantly. The improved anisotropy ({mnl}/{100}) at high temperature provides better conformity to the mask profile for the formation of a micro cavity. The activation energy of TMAH + Triton (0.1-0.2 eV) is lower than that of pure TMAH (or KOH) solution (0.5-0.7 eV), showing to some extent diffusion-controlled etching process.

Secondly, the underlying effect of the surfactant in etching is understood microscopically and proved macroscopically that enables manufacturing of advanced and exciting structures for MEMS. Thicknesses of surfactant layers are investigated depending on the variation of orientation, temperature et al. The pre-adsorbed surfactant layers are formed and their effects on etch rate, surface roughness and corner undercutting indicate that the dissolved surfactant is adsorbed on the surface during etching.

Finally three applications by using surfactant-modified etching process are exhibited, involving the fabrication of conformal structures, scalloping removing for vertical micro-lens, and sharp tips with high aspect ratio. Much effort will be dedicated on other potential aspects in MEMS, and more advanced devices made by this etching technique could be anticipated in future.

6. Acknowledgment

We acknowledge support by MEXT (Micro/Nano Mechatronics G-COE and grant-in-aid for scientific research (A) 19201026 and 70008053) and the Chinese High-level University Program.

7. References

Bütefisch, S.; Schoft, A. & Büttgenbach, S. Three-Axes monolithic silicon low-g accelerometer. *Journal of Microelectromechanical Systems*, Vol.9, No.4, (December 2000), pp. 551-556, ISSN 1057-7157

Gennissen, P, T, J. & French, P, J. Sacrificial oxide etching compatible with aluminium metallization, *Proceedings of IEEE 1997 9th International Conference on Solid-State Sensors and Actuators*, pp. 225–228, ISBN 0-7803-3829-4, Chicago, Illinois, USA, June 16-19, 1997

Gosalvez, M, A.; Pal, P.; Tang, B. & Sato, K. Atomistic mechanism for the macroscopic effects induced by small additions of surfactants to alkaline etching solutions. *Sensors and Actuators A*, Vol.157, No.1, (January 2010), pp. 91-95, ISSN 0924-4247

Gosalvez, M, A.; Tang, B.; Pal, P.; Sato, K.; Kimura Y. & Ishibashi, K. Orientation and concentration dependent surfactant adsorption on silicon in aqueous alkaline solutions: explaining the changes in the etch rate, roughness and undercutting for MEMS applications. *Journal of Micromechanics and Microengineering*, Vol.19, No.12, (December 2009), pp. 125011, ISSN 1361-6439

Hoffmann, M. & Voges, E. Bulk silicon micromachining for MEMS in optical communication systems. *Journal of Micromechanics and Microengineering*, Vol.12, No.4, (July 2002), pp. 349-360, ISSN 1361-6439

Inagaki, N. ; Sasaki, H. ; Shikida, M. & Sato, K. Selective removal of micro-corrugation by anisotropic wet etching, *Proceedings of IEEE 2009 15th International Conference on Solid-State Sensors, Actuators and Microsystems*, pp. 751-754, ISBN 978-1-4244-4193-8, Denver, Colorado, USA, June 21-25, 2009

Lee, K, N.; Lee, D, S.; Jung, S, W.; Jang, Y, H.; Kim Y, K. & Seong, W, K. A high-temperature MEMS heater using suspended silicon structures. *Journal of Micromechanics and Microengineering*, Vol.19, No.11, (November 2009), pp. 115011, ISSN 1361-6439

Ohara, J. ; Kano, K. & Takeuchi, Y. Development of fabrication process for integrated micro-optical elements on Si substrate. *Sensors and Actuators A*, Vol.143, No.1, (May 2008), pp. 77-83, ISSN 0924-4247

Pal, P. & Sato, K. Various shapes of silicon freestanding microfluidic channels and microstructures in one step lithography. *Journal of Micromechanics and Microengineering*, Vol.19, No.5, (May 2009), pp. 055003, ISSN 1361-6439

Pal, P.; Sato, K.; Gosalvez, M, A.; Kimura, Y.; Ishibashi, K.; Niwano, M.; Hida, H.; Tang, B. & Itoh, S. Surfactant adsorption on single crystal silicon surfaces in TMAH solution: Orientation-dependent adsorption detected by in-situ infra-red spectroscopy, *Journal of Microelectromechanical Systems*, Vol.18, No.6, (December 2009), pp. 1345-1356, ISSN 1057-7157

Pal, P.; Sato, K.; Gosalvez, M, A.; Tang, B. & Hida, H. Advanced MEMS applications using orientation dependent adsorption of surfactant molecules in TMAH solution, *Proceedings of 2010 5th Asia-Pacific Conference on Transducers and Micro-Nano Technology*, Perth, Australia, July 6-9, 2010

Park, J.; Park, K.; Choi, B.; Koo, K.; Paik, S.; Park, S.; Kim, J. & Dan Cho, D. A novel fabrication process for ultra-sharp, high-aspect ratio nano tips using (111) single crystalline silicon, *Proceedings of IEEE 2003 12th International Conference on Solid-State Sensors, Actuators and Microsystems*, pp. 1144-1145, ISBN 0-7803-7731-1, Boston, Massachusetts, USA, June 8-12, 2003

Petersen, K, E. Silicon as a mechanical material. *Proceedings of IEEE*, Vol.70, No.5, (May 1982), pp. 420–457, ISSN 0018-9219

Resnik, D.; Vrtacnik, D.; Aljancic, U.; Mozek, M. & Amon, S. The role of Triton surfactant in anisotropic etching of {110} reflective planes on (1 0 0) silicon. *Journal of Micromechanics and Microengineering*, Vol.15, No.6, (June 2005), pp. 1174-1183, ISSN 1361-6439

Sarro, P, M.; Brida, D.; Vlist, W, V, D. & Brida, S. Effect of surfactant on surface quality of silicon microstructures etched in saturated TMAHW solutions. *Sensors and Actuators A*, Vol.85, No.1-3, (August 2000), pp. 340-345, ISSN 0924-4247

Sato, K.; Shikida, M.; Yamashiro, T.; Asaumi, K.; Iriye, Y. & Yamamoto, M. Anisotropic etching rates of single-crystal silicon for TMAH water solution as a function of crystallographic orientation. *Sensors and Actuators A*, Vol.73, No.1-2, (October 1998), pp. 131-137, ISSN 0924-4247

Sato, K.; Uchikawa, D. & Shikida, M. Change in orientation–dependent etching properties of single–crystal silicon caused by a surfactant added to TMAH solution. *Sensors and Materials*, Vol.13, No.5, (May 2001) pp. 285–291, ISSN 0914- 4935

Seidel, H.; Csepregi, L.; Heuberger, A. & Baumgartel, H. Anisotropic etching of crystalline silicon in alkaline solutions: I. Orientation dependence and behavior of passivation layers. *Journal of The Electrochemical Society*, Vol.137, No.11, (November 1990), pp. 3612–3626, ISSN 1945-7111

Shikida, M. ; Sato, K. ; Tokoro, K. & Uchikawa, D. Difference in anisotropic etching properties of KOH and TMAH solutions. *Sensors and Actuators A*, Vol.80, No.2, (March 2000), pp. 179–188, ISSN 0924-4247

Shikida, M.; Hasada, T. & Sato, K. Fabrication of a hollow needle structure by dicing, wet etching and metal deposition. *Journal of Micromechanics and Microengineering*, Vol.16, No.10, (October 2006), pp. 2230-2239, ISSN 1361-6439

Tabata, O.; Asahi, R.; Funabashi, H.; Shimaoka, K. & Sugiyama, S. Anisotroplc etching of silicon in TMAH solutions. *Sensors and Actuators A*, Vol.34, No.1, (July 1992), pp. 51-57, ISSN 0924-4247

Tanaka, H.; Abe, Y.; Inoue, K.; Shikida, M. & Sato, K. Effects of ppb–level metal impurities in aqueous potassium hydroxide solution on the etching of Si{110} and {100}. *Sensors and Materials*, Vol.15, No.1, (January 2003) pp. 43–51, ISSN 0914- 4935

Tang, B.; Amakawa, H.; Shikida, M.; Hida, H; Inagaki, N.; Pal, P. & Sato, K. Characterization of etching anisotropy in a surfactant-added TMAH solution, and its application to scalloping reduction, *Proceedings of 2010 5th Asia-Pacific Conference on Transducers and Micro-Nano Technology*, Perth, Australia, July 6-9, 2010

Tang, B.; Gosalvez, M, A.; Pal, P.; Itoh, S.; Hida, H.; Shikida, M. & Sato, K. Adsorbed surfactant thickness on a Si wafer dominating etching properties of TMAH solution, *Proceedings of IEEE 2009 20th International Symposium on Micro-Nano Mechatronics and Human Science*, pp. 48-52, ISBN 978-1-4244-5094-7, Nagoya, Japan, Nov 9-11, 2009

Tang, B.; Pal, P.; Gosalvez, M, A.; Shikida, M.; Sato, K.; Amakawa, H. & Itoh, S. Ellipsometry study of the adsorbed surfactant thickness on Si{110} and Si{100} and the effect of pre-adsorbed surfactant layer on etching characteristics in TMAH. *Sensors and Actuators A*, Vol.156, No.2, (December 2009), pp. 334-341, ISSN 0924-4247

Tang, B.; Sato, K.; Tanaka H. & Gosalvez, M, A. Fabrication of sharp tips with high aspect ratio by surfactant-modified wet etching for the AFM probe, *Proceedings of IEEE 2011 24th International Conference on Micro Electro Mechanical Systems*, pp. 328-331, ISBN 978-1-4244-9632-7, Cancun, Mexico, January 23-27, 2011

Tang, B.; Shikida, M.; Sato, K.; Pal, P.; Amakawa, H.; Hida, H. & Fukuzawa, K. Study of surfactant-added TMAH for applications in DRIE and wet etching-based micromachining. *Journal of Micromechanics and Microengineering*, Vol.20, No.6, (June 2010), pp. 065008, ISSN 1361-6439

Yang, C, R.; Chen, P, Y.; Yang, C, H.; Chiou, Y, C. & Lee, R, T. Effects of various ion–typed surfactants on silicon anisotropic etching properties in KOH and TMAH solutions. *Sensors and Actuators A*, Vol.119, No.1, (March 2005), pp. 271-281, ISSN 0924-4247

Yang, C, R.; Yang, C, H. & Chen, P, Y. Study on anisotropic silicon etching characteristics in various surfactant-added tetramethyl ammonium hydroxide water solutions. *Journal of Micromechanics and Microengineering*, Vol.15, No.11, (November 2005), pp. 2028-2037, ISSN 1361-6439

Zubel, I. & Kramkowska, M. The effect of isopropyl alcohol on etching rate and roughness of (100) Si surface etched in KOH and TMAH solutions. *Sensors and Actuators A*, Vol.93, No.2, (September 2001), pp. 138-147, ISSN 0924-4247

8

Macromodels of Micro-Electro-Mechanical Systems (MEMS)

Anatoly Petrenko

Systems Design Department, National Technical University of Ukraine
Ukraine

1. Introduction

Micro-electro-mechanical Systems (MEMS) are components with micron-scale moving parts based on materials and processes of microelectronics fabrication. This is a good example of on-chip integration of electronics, microstructures, microsensors and microactuators. Accurate simulation of MEMS requires precise modeling of all effects of mechanical and damping forces, electrostatic forces and inner stresses, heat transfer, thermal expansion, piezoelectric stresses etc.

Modern methodology of MEMS design implies that the entire MEMS can be investigated only at higher abstraction levels such as **schematic and system** ones, where accurate macromodels can be used [1]. On the other hand, at component or device levels the physical behavior of three-dimensional continuums is described by **partial differential equations** (PDE) easily solvable by Finite Element or Finite Difference Element Methods (FEM or FDM) [2,3], available in ANSYS –like software. Component level simulations are classified in single - domain and coupled - domain simulations, both being very computer time- consuming.

The goal of this chapter is to consider methods of automatically obtaining macromodels of MEMS and their mechanical or non-electric components from ANSYS models as equivalent electric circuits or low order differential ordinary equations for further use in circuit design software. This can be done by using different model order reduction techniques developed in recent years.

When dealing with the modern MEMS, the possibility for using a single environment to simulate objects, where different physical processes such as electrical, mechanical, optical, thermal etc. take place, plays an important role. Here we have to represent different subsystems of the initial MEMS as equivalent models of the same physical nature permitting to combine them for solution in a single computational process. After that, the complete behavioral model of the entire MEMS and its subsystems can be compiled either in VHDL-AMS language (as sets of ODE) or in SPICE-like language (as equivalent electric circuits).

The Microsystems design exploits various analytical and numerical methods for virtual prototyping of MEMS. It also demands for libraries of electromechanical, optical models and microfluid components, including springs, bulks, buffers, capacitors, inductances, operational amplifiers, transistors etc. Three basic possible approaches of MEMS design procedure are illustrated below: FEM/FDM Model, Reduced Order Model (ROM), Coupled system-level model.

2. FEM/FDM model

MEMS typically involve multiple energy domains such as kinetic energy, elastic deformation, electrostatic or magneto static stored energy and fluidic interactions. The difficulty in the modeling of MEMS devices is mainly caused by the tight coupling between the multiple energy domains. Individual physical effects are governed by partial differential equations (PDE), typically nonlinear. When these equations become coupled, the computational challenges of highly meshed numerical simulation become formidable.

FEM relies on highly localized interpolation functions (or *mesh element functions*) for approximation of the solution of PDE. These mesh element functions are generated by meshing the domain of interest and parameterize the desired solution locally on each mesh element. This parameterized solution converts a continuous (PDE) problem to a coupled system of ordinary differential equations (ODE) that can be integrated in time. The resulting ODE system usually has many degrees of freedom (perhaps **several variables per mesh element**). If a fine mesh is required, the problem size grows rapidly, with a corresponding rapid growth in computational cost for explicit dynamic simulation. Consequently, it is very expensive to use FEM model in system-level simulations during MEMS iterative design. As a result, FEM models are mostly used to analyze the performance of MEMS components and to couple their multiphysics effects.

By reading ANSYS binary FULL file it is possible to assemble a MEMS component state-space model in the form of first order systems or second order ordinary differential equations (ODE)

$$E_r z' + A_r z = B_r f , \qquad Y = C_r z \tag{1}$$

$$Mx'' + Dx' + Kx = Bf , \quad Y = Q^T x + R^T x' , \tag{2}$$

where A_r, E_r, C_r, B_r, M, D, K, B, C- are the system matrices, B_r, B are the input and the C_r, C - output matrices, f is input force. In mechanics matrices M, D and K are known as the *mass*, *damping* and *stiffness* matrices correspondingly. Usually damping is included in the model as Rayleigh damping. The damping matrix D is computed as a linear combination of the stiffness K and the mass M matrices:

$$D = \alpha M + \beta K,$$

where α, β are constant coefficients.

In (1) the state space vector z is defined through the unknowns deflections $u(x,t)$ and pressures $p(x,y,t)$ into the node points being automatically generated in MEMS structure:

$$z = [u_1 ... u_N \frac{\partial u_1}{\partial t} ... \frac{\partial u_N}{\partial t} p_{11} ... p_{MN}]^T \tag{3}$$

By defining

$$E_r = \begin{bmatrix} D & M \\ M & 0 \end{bmatrix} \quad A_r = \begin{bmatrix} K & 0 \\ 0 & -M \end{bmatrix} \quad B_r = \begin{vmatrix} B \\ 0 \end{vmatrix} \quad C_r = \begin{vmatrix} Q \\ R \end{vmatrix} \quad z = \begin{vmatrix} x' \\ x'' \end{vmatrix} \tag{4}$$

second equations (2) can be transfer to the first (1).

The FULL file contains all the information about the system: the system element matrices, Dirichlet boundary conditions, equation constrain and the load vector. It is generated using ANSYS partial solver, which enables to assemble system element matrices for the desired analysis without solving them and it therefore computationally fast. The speed of the reading operation has been optimized taking into account that the element matrices are sparse. The load vector directly gives the matrix-vector product Bf and thus describes the distribution of all loads being applied. In order to obtain the B matrix, and thus being able to modify the inputs singularly, it is necessary to repeat the partial solution for each input of interest.

3. ROM (Reduced Order Model)

It would be easier and more intuitive for the designer to explore the design space if the MEMS model had only a few variables with a clear relationship between them and the overall device performance. *Reduced-order models (ROM)*, also called *macromodels*, lend themselves very well to these purposes. The main idea behind the reduced order model is that the number of ordinary differential equations (ODE) needed to simulate the system has been reduced from perhaps many thousands in the case of the full FEM simulation, to just a few basis function coordinates (fig.1).

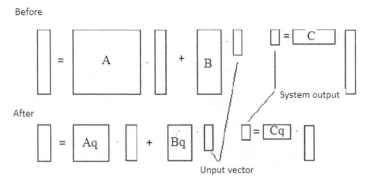

Fig. 1. Reduced order model illustration

Such the macromodel simulation can be very efficient computationally compared to the FEM model. A designer can use the FEM model for different component geometry and materials trying and the ROM model for investigation of different input forces effect (fig.2).

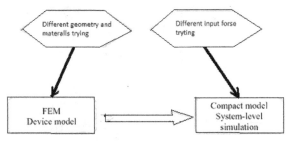

Fig. 2. Compact reduced order model in MEMS design [33]

3.1 Modal decomposition ROM

Modes (or resonances) are inherent properties of a structure. Resonances are determined by the material properties (mass, stiffness, and damping properties) and boundary conditions of the structure. Each mode is defined by a *natural (modal or resonant) frequency, modal damping*, and a *mode shape*.

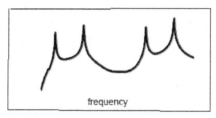

Fig. 3. Simple Plate Frequency Response Function

If either the material properties or the boundary conditions of a structure change, its modes will change as well. The overall response of a structure at any frequency is a *summation of responses due to each of its modes*. It is also evident that close to the frequency of one of the resonance peaks, the response of *one mode will dominate the frequency response* [6], fig.3.

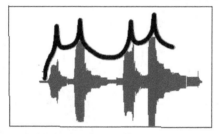

Fig. 4. Overlay of Frequency and Time Response Functions

Now if we overlay the time trace with the frequency trace it is possible to notice that maximum oscillations at the time are corresponding to the frequencies of maximum frequency response function (fig.4).

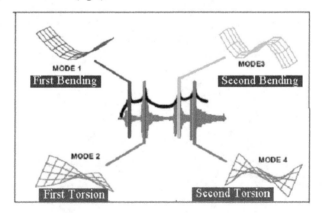

Fig. 5. Flexible Body Modes [6].

Let's see what happens to the deformation pattern on the structure at each one of these natural frequencies. Modes are further characterized as either *rigid body* or *flexible body* modes. All structures can have up to *six rigid body* modes, three translational modes and three rotational modes. Many deformation problems are caused, or at least amplified by the excitation of one or more flexible body modes. Fig.5 shows some of the common fundamental (low frequency) modes of a plate. The fundamental modes are given names like those shown in Fig.5. The higher frequency mode shapes are usually more complex in appearance, and therefore don't have common names.

Modal decomposition method uses a weighted sum of n mode shapes (modal amplitudes or eigenvalues) q_i, basis function (an eigenvector) φ_i (x, y, z)) of the mechanical structure to represent its deflection u:

$$u(x,y,z,t) = u_{eq} + \sum_{i=1}^{n} q_i(t)\varphi_i(x,y,z) \ , \tag{5}$$

where u_{eq} is the initial displacement produced by the initial load. For MEMS it is usually sufficient that few modes accurately describe dynamical response of the system. This approach is equivalent to the projection of the original PDE, describing the MEMS behavior, on the subspace defined by the basis functions.

By inserting (5) into the equation of motion (2), taking $\mathbf{x} = \varphi \, e^{\lambda t}$ and multiplying by φ^T_i from the left, and using the orthogonality of eigenvectors, the equation (2) can be reduced to

$$\ddot{q}_i - 2\sigma_i q_i + (\omega_i^2 + \sigma_i^2) = \varphi_i^T f \ , \quad i=1,2,\ldots n \tag{6}$$

The eigenvectors satisfy the orthogonality conditions $\varphi^T_k M\varphi_i = \varphi^T_k D\varphi_i = \varphi^T_k K\varphi_i = 0$ for $\lambda_k \neq \lambda_i$. With the normalization of the eigenvectors $\varphi^T_k M\varphi_k \equiv 1$, it can be shown that

$$\varphi^T_k D\varphi_k = -2\sigma_k \text{ and } \varphi^T_k K\varphi_k = \omega^2_k + \sigma^2_k \text{ if } \lambda_k = \sigma_k + j\omega_k.$$

Note that the orthogonality condition does not apply in the case of multiple eigenvalues. In such case, special modal analysis techniques are needed for decoupling of modes [7].

The decoupling of modes yields that transfer functions can be written as a sum of *modal transfer functions*. Using Fourier transforming expansion and equations (6), the transfer matrix can be derived as

$$H(\omega) = \sum_{i=1}^{n} H_k(\omega) = \sum_{i=1}^{n} \frac{\varphi_i \varphi_i^T}{(j\omega - \sigma_i - j\omega_i)(j\omega - \sigma_i + j\omega_i)} \tag{7}$$

This relation is the basis of modal analysis. It relates the *measurable* transfer functions to the modal properties ω_i, σ_i, and φ_i. Each i- th mode contributes with a modal transfer matrix H_i to the complete transfer matrix. Fig.6 shows an example of a theoretical transfer function (7) with three *modal peaks* corresponding to modes at 1, 2, and 4 Hz which are all damped with a logarithmic decrement of 1% ($-\sigma_i/f_i = 0.01$). Also the individual modal transfer functions (7) are plotted from which the complete transfer function is computed.

By the way the eigenvalues q_k and the corresponding eigenvectors φ_k for $k = 1, 2, \ldots, n$ can be found as a solution of the modified eigenvalue problem

$$(M\lambda^2 + D\lambda + K)\varphi = 0 \ . \tag{8}$$

The relationships between natural frequencies f_k, logarithmic decrements δ_k and the eigenvalues are $f_k = 2\pi\omega_k$ and $\delta_k = -\sigma_k/f_k$.

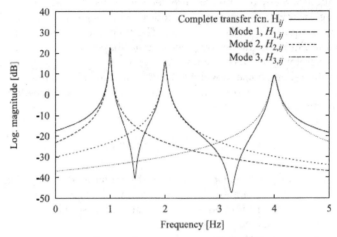

Fig. 6. Example of the modal decoupling in a transfer function [7].

3.1.1 Modal ROM based on proper orthogonal decomposition methods

Reduced order modeling using modal basis functions was originally developed by [11-13] and has been continuously improved by several authors. The basis functions $\varphi_i(x,y,z)$ in (5) may be chosen in two ways by:

- Using the undamped linear mode shapes of the undeflected microstructure as basis functions. For simple structures with simple boundary conditions, the mode shapes can be found analytically. For complex structures or complex boundary conditions, the linear mode shapes are obtained numerically using the finite element method. However, it is usually difficult to determine, *a priori*, an optimum set of eigenvectors φ_k, particularly when irregular geometries are involved.

- Using snapshots obtained from experiments under a training signal or full FEM model runs. Then the proper orthogonal decomposition method (Singular Value Decomposition -*SVD*, Karhunen -Loève decomposition -*KL* and neural networks-based generalized Hebbian algorithm -*GHA*) is applied to the time series for extracting the mode shapes of the device structural elements.

The choice of orthogonal basis functions φ_k can be done by the following way [8]. First the MEMS dynamics are simulated using a slow but accurate technique such as FEM or FDM. Sets of runs may be used to suitably characterize the operating range of the device. The spatial distributions of each state variable $u(x, y, z, t)$ are then sampled at a series of N_s different times during these simulations, and the sampled distributions are stored as a series of vectors, $\{u_i\}$, where each of them corresponds to a particular "snapshot" in time. Now suppose we would like to pick orthogonal basic n functions $\{\varphi_1, \varphi_2,..., \varphi_n\}$ in order to represent the observed state distributions as closely as possible. One way to do this is to attempt to minimize a least squares measure of the "error" distances between the observed states and the basis function representation of those states:

$$\sum_{i=1}^{n} [u_i - proj(u_i, span\{\varphi_1, ..., \varphi_n\})]^2 = \sum_{i=1}^{n} \sum_{j=1}^{n} \varphi_i \varphi_j^T u_i , \tag{9}$$

where $proj(v,S)$ is the projection of the vector v into subspace S. In other words, we minimize a least squares measure of the "error" distances between the observed states and the basis function representation of those states.

Singular value decomposition (SVD) takes a rectangular matrix of modal experimental data (defined as U, where U is a N x n matrix) in which N is a number spatial points of each snapshot, n is a number of snapshots. The SVD theorem states:

$$U_{Nxn} = W_{NxN} \, S_{Nxn} \, V^T_{nxn} ,$$

where the columns of W are the left singular vectors (*spatial points vectors*); S (the same dimensions as U) has singular values and is diagonal (*mode amplitudes*); and V^T has rows that are the right singular vectors (*snapshots numbers vectors*).

The SVD represents an expansion of the original data in a coordinate system where the covariance matrix is diagonal. Calculating the SVD consists of finding the eigenvalues and eigenvectors of matrices UU^T [NxN] and U^TU [n x n]. The eigenvectors of U^TU make up the columns of V, the eigenvectors of UU^T make up the columns of W. Also, the singular values in S are square roots of eigenvalues from UU^T or U^TU. The singular values are the diagonal entries of the S matrix and are arranged in descending order: $\sigma_1 \geq \sigma_2 \geq ... \geq \sigma_n \geq 0$. The singular values are always real numbers. If the matrix U is a real matrix, then W and V are also real.

It was shown [11] that the proper orthogonal basis functions $\{\varphi_1, \varphi_2, ..., \varphi_n\}$ minimizing (9) can be chosen by setting

$$\phi_i = v_i, i = \{1, 2, ..., n\}$$

where v_i is the columns of V.

After finding basis functions φ_k the linear equations for modal amplitudes q_k estimation are formed for each temporal snapshot on the base of selected before "snapshot" spatial points and equation (5).

Karhunen-Loève decomposition (KL) can be viewed as a statistical procedure [9]. One initially supposes that the observed system dynamics can be modeled as a second-order ergodic stochastic process. The method consists then in constructing a spatial autocorrelation tensor from data obtained through numerical or physical experiments and performing its spectral decomposition.

KL differs from SVD in way of finding basis functions φ_i through the temporal snapshot

correlation matrix entries
$$a_{ij} = \frac{1}{n} \sum_{\substack{k=1 \\ m=1}}^{n} [u_i(x,t_m)u_j(x,t_k)]$$

as
$$\varphi_i(x) = \sum_{k=1}^{n} b_{k,i} u_k(x) ,$$

where n is number of temporal snapshots and $b_{k,i}$ are eigenvectors of the matrix A, or the solutions of the equation $Ab = \lambda b$.

Generalized Hebbian algorithm (GHA) is an Artificial Neural Network (ANN) approach of performing Principal Component Analysis (PCA) on a set of data and can be also used as a learning procedure for the approximation of *PDE* solutions by the expression (5) [10] .

The modal approach to MEMS macromodeling is illustrated by sensor device (fig.7), which is described by coupling a 1-D elastic beam equation with electrostatic force and 2-D compressible isothermal squeeze-film Reynold's equation [11]. This device consists of a deformable elastic beam microstructure that is electrostatically pulled in by an applied voltage waveform. The dynamics of the beam are first simulated using a finite-difference analysis. A quarter of the beam is initially meshed using a 20 x10 node 2-D grid.

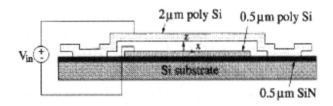

Fig. 7. Fixed-fixed beam pull-in time pressure sensor device [11]

The state at each node consists of three quantities: z, dz/dt and p. Since z and dz/dt are simulated in 1-D, this results in coupled nonlinear ODE which must be integrated in time. Basis functions are generated for pressure p and displacement z based on runs of the finite-difference code for an ensemble of four different step voltages: 9V, 10V, 12V, 16V. One hundred samples of pressure and displacement are taken during these four runs at fixed time intervals. These samples are used to generate the basic functions. The resulting basis functions for displacement and pressure are shown in Fig.8 [11].

3.1.2 Nonlinear modal ROM

The device governing equations are generally derived from Lagrange equations, after expressing the internal (elastic and kinetic) and external (electrostatic) energy of the system in terms of modal amplitudes and symbolical calculation of the gradients. Assuming that the device undergoes small displacement, the basis chosen results in diagonal mass and stiffness matrices, which can be pre-computed. Linear elastic undamped normal modes of the undeflected device have been often chosen as basic functions to approximate the solution of an electromechanical problem discretized using finite element methods. Modal representation is very efficient since it *requires only one equation per mode* and involved conductor to describe the entire system.

The nonlinear energy terms, instead, are generally expressed as analytical functions of the modal coordinate. In [13] a single static full finite element simulation is used to determine the number of modal functions needed to capture the device behavior and their expected amplitude. Then this information is used to construct the electrostatic energy term. A 3D full model electrostatic simulation is run for values of the modal amplitude that span the operating range of the device, in order to compute the capacitance/deformation curve by using ANSYS' transducer element TRANS126. The results are fitted with a rational fraction of multivariate polynomials using a nonlinear function fitting scheme. In order to model large-displacement behavior of the device, the strain energy is derived by fitting data from a set of full finite element simulations [13]. A similar procedure is proposed in [17]. In this case, the force-displacement function and the modal strain energy are still derived from a series of FEM simulations, but a polynomial multi-variable fit is used. In order to reduce the complexity of the fitting step, dominant and relevant modes are first characterized. Both the procedures in [13] and [17] can be partially automated and extended to include other conservative energy domains. Other algorithms have also been proposed for the approximation of dissipative energy terms [18, 19]. Dedicated methods have been demonstrated for actuated microbeams, still using the linear undamped mode shapes of the device as basic functions in the Galerkin procedure [20], where two expressions of the nonlinear electrostatic term were proposed as a function of modal coordinates, each including all nonlinearities up to the fifth order, obtained via mathematical manipulation [21]. A new fuzzy-logic model (FLM) for MEMS is presented in [22] in which for reducing the number of data needed for macromodel identification, cluster estimation of a model structure and back propagation method of structure parameters adaptation are chosen to fit the data. As a result the dynamic coupled simulation of a magnetic microactuator takes only several minutes and the force macromodel yielded errors is less than 1.5% for a 5-μm displacement. These FLMs combine fuzzy sets with fuzzy rules that have the capability to model the complex nonlinear behavior.

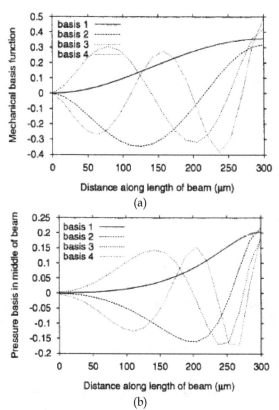

Fig. 8. Basis functions for (a) displacement $z(x, t)$ and (b) pressure $p(x, y, t)$ [11]

Macromodels obtained via modal basis functions methods have been demonstrated to reproduce results obtained with full physical level simulation with an accuracy of some percentage points and a reduction of the computational complexity simulation. It was shown that even when the problem is mechanically nonlinear, the linear normal modes can serve as basic functions. The approach can be outlined as follows [13]:

1. Compute the linear modes φ_i of the elastic problem.
2. Substitute u (x, y, z, t) in the governing equation for the deflection (e.g., the Euler–Bernoulli or Timoshenko equations for beams).
3. Obtain a system of n-coupled second-order ordinary differential equations for the $q_i(t)$.
4. Solve the equations to compute the dynamic response either numerically or analytically.

3.1.3 VHDL-AMS export of modal ROM

In general, the equation (5) describes a coordinate transformation of finite element displacement coordinates (mesh element coordinate) to modal coordinates of the macromodel (basic functions or a degree of freedom-d.o.f.):

$$u(x,y,z,t) = u_{eq} + \sum_{i=1}^{n} q_i(t)\varphi_i(x,y,z) .$$

The deformation state of the structure given by N nodal displacements u_i $(i=1,2,...,N)$ is now represented by a linear combination of n modes weighted by their amplitudes q_j $(j=1,2,...,n)$ where $n \ll N$.

The governing equation of motion describing the ROM of electrostatic actuated MEMS structures in modal coordinates:

$$m_j\ddot{q}_j + 2\xi_j\omega_jm_j\dot{q}_j + \frac{\partial}{\partial q_j}W_{st}(q_1,...,q_n) = \frac{1}{2}\sum_r\frac{\partial}{\partial q_j}C_{ks}(q_1,...,q_n)\cdot(V_k - V_s)^2 + \sum_{i=1}^{n}\varphi_if_i , \qquad (10)$$

where m_j is the modal mass, ω_j is the eigenfrequency, ξ_j the linear modal damping ratio, W_{st} is the modal strain energy function, C_{ks} is the modal capacity-stroke function, r is the number of capacities involved for microsystems with multiple electrodes, V is the electrode voltage applied and f_i is a local force acting at the i-th node. The current I_k at each electrode k is defined by:

$$I_k = \frac{\partial Q_k}{\partial t} = \sum_r(C_{ks}\cdot(\frac{\partial V_k}{\partial t} - \frac{\partial V_s}{\partial t}) + \frac{\partial C_{ks}}{\partial t}\cdot(V_k - V_s)) \qquad (11)$$

An essential prerequisite to establish (10) and (11) are proper modal strain energy and capacity-stroke functions. Both are derived from a series of FEM runs at various deflection states in the operating range. The received data are used for polynomial functions fitting in order to compute the local derivatives, which describe force and stiffness terms. As a matter of fact, shape function methods can be applied to nonlinear systems, too [13]. Geometric nonlinearities, as, for instance, stress stiffening, can be regarded if the modal stiffness is computed from the first derivative of the strain energy function with respect to the modal amplitudes. Capacitance-stroke functions provide non-linear coupling between each eigenmode and the electrical quantities (i.e. electrostatic modal forces, electrical current) if stroke is understood as modal amplitude. Damping parameters are assigned to each eigenmode.

The first step of the ROM generation is to determine which modes are really significant, and to estimate a proper amplitude range for each mode. Several criteria can be applied, for instance, the lowest eigenmodes of a modal analysis, modes in operating direction, or modes, which contribute to the deflection state at a typical test load. Next the dependencies of the strain energy W_{st} and capacities are described by polynomial functions being fitted. The necessary data points are obtained by imposing each eigenmode with varying amplitude on the mechanical model1 for the non-linear strain energy and on an electrostatic space model for capacitance. This process is computationally expensive but has to be done just once. The result is a black-box model that can be applied to any load situation. In the concept of the modal superposition method, each eigenmode represents a single independent resonator with modal mass m_i and modal damping ξ_i.

The export of the ROM to VHDL-AMS is performed in two steps [16]. At first, an initialization file containing all necessary information of the macromodel, such as the fitted polynomial coefficients and orders, is generated. Then, the source code in VHDL-AMS is automatically generated. The main problem of exporting the ROM in VHDL-AMS is to express the fitted functions of the non-linear strain energy and of the capacities which are part of coefficients of the differential algebraic equations (DAE), which can be mapped to the simultaneous statements of VHDL-AMS. If simulators support description in matrix notation properly (as MATLAB), the exported VHDL-Models will become more compact and clear.

The Modal ROM approach was implemented as the available ROM-Tool in ANSYS/Multiphysics since Release 7 **(ROM144).** It contains some terminals (fig. 9) and provides necessary functions:

- the master node terminals which describe the displacement u_i and the inserted forces $F_{N,i}$ at these nodes;
- the modal terminals with the modal amplitude q_i and modal force $F_{M,i}$ for the chosen modes;
- the electrical terminals which provide the voltages V_i and currents I_i for the electrodes of the system.

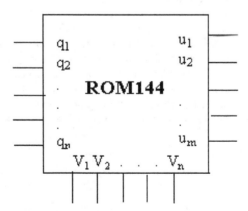

Fig. 9. ROM144 functional block

The ROM144 takes input equations (5),(10),(11) with characteristic constants (modal masses, modal damping ratios and eigenvectors of the master nodes) and provides the special functions of calculating the strain energy W_{mech} (q_i, q_j , q_k) and capacitances $C_{op}(q_i, q_j , q_k)$ as well as their first derivatives with respect to the modal amplitudes q_i, q_j and q_k using the information of the polynomials degrees defined in other packages.

Modal ROM144 approach speeds up computations in 40 times in comparison with the FEM model while pull-in time errors is less than 2%.

Modal ROM approach was implemented also in *INTEGRATOR system of CoventorWare* (http:// www.coventor.com), *MEMS Pro* (http://www.memspo.com) and *MEMSCAP* (http:// www.memsscap.com).

3.1.4 Galerkin's approximated ROM

As in (5) the desired PDE solution $u(x,y,z,t)$ can be approximated by spatially varying arbitral basis functions $\beta_i(x,y,z)$ with time varying coefficients $\alpha_i(t)$:

$$u(x,y,z,t) = \sum_{i=1}^{n} \alpha_i(t)\beta_i(x,y,z) \tag{12}$$

For the Galerkin's method the PDE residual $(L(u)-f)$ is orthogonal to each a_i of the basic functions in the operating range H:

$$(\alpha_i, L(u) - f) = \int \alpha_i^T (L(u) - f)dt = 0, \quad i = 1,n. \tag{13}$$

where L is a differential operator (possibly nonlinear), and f is an input vector.

The basic functions $\beta_i(x,y,z)$ can be chosen arbitrarily, as long as their elements satisfy all of the boundary conditions and are sufficiently differentiable. It means that they can be not only eigenmodes as it was shown before, but they may be Tchebychev, Legendre, Hermit polynomials or even wavelets functions, which were introduced in the past two decades and are gaining increasing popularity. Indeed wavelets have many excellent properties: such as orthogonality, compact support, exact representation of polynomials to a certain degree, and flexibility to represent functions at different resolution levels. The wavelet-Galerkin method is a Galerkin scheme using wavelet functions as the basic functions. However, wavelet functions do not satisfy the boundary conditions. Thus the treatment of general boundary conditions is a major difficulty for the application of the wavelet- Galerkin method. The wavelet interpolation Galerkin method is a Galerkin scheme where basic functions are a class of interpolating functions generated by autocorrelation of the usual compactly supported Daubechies scaling functions [23]. Daubechies' functions are easy to construct. For an even integer L, we have the Daubechies' scaling function $\varphi(x)$ and wavelet $\psi(x)$ satisfying

$$\varphi(x) = \sum_{i=1}^{L-1} p_i \varphi(2x - i)$$

$$\psi(x) = \sum_{i=2-L}^{1} (-1)^i p_{1-i} \varphi(2x - i) \tag{14}$$

The scaling function $\varphi(x)$ is supported in the interval $[0, L-1]$ while the corresponding wavelet $\psi(x)$ is supported in the interval $[1 - L/2, L/2]$. The parameter L will be referred to as the degree of the scaling function $\varphi(x)$. The coefficients p_i are called the wavelet filter coefficients and they satisfy the same conditions. The constructed scaling function $\varphi(x)$ and wavelet $\psi(x)$ have also the prescribed properties.

The autocorrelation functions $\theta(x)$, which are used for generating basic functions, can be defined as follows:

$$\theta(x) = \int_{-\infty}^{\infty} \varphi(\tau)\varphi(\tau - x)d\tau \tag{15}$$

and act as the scaling function

$$\theta_{J,k}(x) = \theta(2^J x - k). \tag{16}$$

Wavelets have proven to be an efficient tool of analysis in many fields including the solution of PDE. In [23] a new wavelet interpolation Galerkin method is used for the numerical simulation of MEMS devices under the effect of squeeze film damping. The air film pressure is expressed as a linear combination of a class of basic functions generated by autocorrelation of the usual compactly supported Daubechies scaling functions, which are the first- generation wavelets. The wavelet interpolation Galerkin method was used to predict the frequency response of the accelerometer with a spring mass-damper model with a parallel-plate electrostatic force. Various numerical tests have been conducted by changing the degree of the Daubechies wavelet L and the number J of the scale. Better accuracy can be achieved by increasing L and J. The higher L is, the smoother the scaling function becomes. The price for the high smoothness is that its supporting domain gets larger. The higher J is, the more accurate the solution becomes. The number of differential equations and the CPU time increase significantly as J increases. The solutions for $L = 6$ and $J = 4$ have results higher than the finite difference method.

The present wavelet interpolation Galerkin method is not suitable to solve problems defined on nonrectangular domains, since higher-dimensional wavelets are constructed by employing the tensor product of the one-dimensional wavelets, so their application is restricted to rectangular domains. But usage of the second-generation wavelets which are constructed in the spatial domain can expend in future PDE solutions to complex domains.

3.2 Moment matching based ROM

Moment matching model order reduction is based on the approximation of the original n-dimensional system transfer function $F(s)$ of the original n-dimensional system with a rational function with a lower degree $q<<n$ [24]. This is done by matching some terms of the Taylor expansion of $F(s)$ around a certain expansion point.

For a state space equation

$$\dot{X}(t) = AX(t) + Bv(t)$$
$$Y(t) = C^T X(t) \tag{17}$$

let's perform a Laplace transform to obtain its frequency domain transfer function

$$F(s) = C^T (sI - A)^{-1} B \tag{18}$$

and expand it into Taylor series as

$$F(s) = -\sum_{k=0}^{\infty} m_k s^k \tag{19}$$

where $m_k = C^T A^{-k}(A^{-1}B)$.

Moment matching is directly connected to the Krylov subspace formed by the pair of matrices $(A^{-1}, A^{-1}B)$ [24]. The Krylov subspace is spanned by the column vectors in the following collection of matrices:

$$\{A^{-1}B, (A)^{-1}A^{-1}B, \ldots, (A)^{-i}A^{-1}B, \ldots\} \tag{20}$$

where the column vectors are called the Krylov vectors. The q-th order Krylov subspace is denoted by

$$K_q(A^{-1}, A^{-1}B) \tag{21}$$

which is spanned by the leading q linearly independent Krylov vectors in (20). Let $V \in R^{n \times q}$ be any matrix which columns span the Krylov subspace $K_q(A^{-1}, A^{-1}B)$

$$\text{Spancolumn}\{V\} = \text{span}\{A^{-1}B; A^{-2}B; \ldots; A^{-q}B\}.$$

Here V is the orthogonal projection matrix that maps the n-dimensional state-space into a q dimensional state-space and satisfies $V^T V = I$. If the columns of V are orthogonal and B is a column vector, it can be shown that the following identities hold [24]:

$$(A)^i A^{-1}B = V(A_q)^i A_q^{-1}B_q \tag{22}$$

for $i = 0, 1, \ldots, q - 1$. These identities can be used to verify that at least the q leading moments of the full-order and reduced-order transfer functions are matched. Finally, we get the reduced order system of much smaller order (or state-space dimension) by performing variable change $x(t) = V\tilde{x}(t)$ and multiplying on V^T both sides of the equations (17):

$$\begin{aligned} \tilde{x}'(t) &= A_q\tilde{x}(t) + B_q v(t) \\ y(t) &= C_q^T \tilde{x}(t) \end{aligned} \tag{23}$$

where $\quad A_q = (V^T A V), \quad B_q = V^T B, \quad C_q = C^T V.$

As a common practice, the block vectors forming the Krylov subspace are orthogonalized by using the Arnoldi algorithm for a numerical stability. Lets describe the block Arnoldi algorithm for a single column input matrix, i.e., $B = b$, where $b \in R^n$, so that the algebraic operations involved can be seen.

3.2.1 Arnoldi Algorithm

i.   LU factorize matrix A: $A = LU$.

ii.   Solve $\tilde{v}_1$ from: $A \tilde{v}_1 = b$.

iii. Compute $h_{11} = \|\tilde{v}_1\|$ and $v_1 = \tilde{v}_1 / h_{11}$.

iv. For $j = 2, \ldots, q$:

Solve $\tilde{v}_j$ from: $A\tilde{v}_j = Av_{j-1}$.

For $i = 1, \ldots, j - 1$: $h_{ij} = v_i^T \tilde{v}_j$.

$$w_j = \tilde{v}_j - \sum_{i=1}^{j-1} v_i h_{ij}$$

$$h_{jj} = \left\| w_j \right\|, \quad v_j = w_j / h_{jj}.$$

Note that the Arnoldi algorithm terminates when $h_{jj} = 0$, which means that the subsequent vectors belong to the subspace already generated. The Arnoldi algorithm is basically a Gram–Schmidt procedure for orthogonalizing the Krylov vectors. The variant of the Arnoldi algorithm (also known as *PRIMA*) with some extra computational effort preserves the passivity of the original system [27].

The Moment matching method of model order reduction can be extended on nonlinear systems. In these cases the original nonlinear systems has to be changed previously (linearized or piecewise-linearized, approximated by a quadratic systems, divided into several linear systems, etc.) [28].

3.2.2 Second order systems

A size of a second order equations system (2) can be also reduced by transforming it to the first order system (1), and then applying the methods described before. However, the reduction of second order systems by such transformation ignores the physical meaning of the original matrices and gives a reduced order model in a first order form. It is desirable for the reduced system to preserve the form of the original system (2). Approaches, that deal directly with the system (2) reduction have been proposed in the framework of Krylov subspaces methods [29, 30].

The transfer function for the system (2), with zero initial conditions, is given by:

$$H(s) = C^T (s^2 M + sD + K)^{-1} B \tag{24}$$

If the system is **undamped**, i.e. $D = 0$, the Arnoldi process can be applied for the computation of a basis for the Krylov subspace K_q $(K^{-1}M, K^{-1}B)$ which is used for the projection matrix V building. The transfer function can be expanded into Taylor series as

$$H(s) = -\sum_{k=0}^{\infty} m_k s^{2k},$$

where $m_k = C^T (K^{-1}M)^{-k}(A^{-1}B)$.

Then the reduced system matrices with variable change $x(t) = V\tilde{x}(t)$ are obtained by the orthogonal projection:

$$M_q = V^T MV, \quad K_q = V^T KV, \quad B_q = V^T B, \quad C_q = C^T V \tag{25}$$

and the matching properties of the method are conserved [31].

It is worth noting that, starting from a second order system in the form (2), Krylov subspace methods require the knowledge of K^{-1}. For high dimensional systems, the explicit calculation of the inverse of K is computationally not affordable. Its computation is therefore replaced by the solution of linear systems of equations through a *LU*- decomposition and defying $K^{-1} = U^{-1} L^{-1}$.

If the system is **damped**, i.e. $D \neq 0$, in case of Rayleigh damping, it demonstrates that the damping matrix can be neglected during the reduction process and it can be computed afterwards, as a linear combination of the reduced stiffness and mass matrices [29]. That is why it is possible to recalculate the matrix D also similar to matrices M and K as in (25):

$$D_q = V^T D V.$$

The reduction of a nonlinear system (2) can be done by using linear model order reduction techniques and considering nonlinearities as inputs. In the general case, the stiffness matrix is nonlinear since its entries dependent on the nodal displacements x. So damping matrix entries also are nonlinear functions of the nodal displacements x and the applied voltages V. If the nonlinearities are confined in the input function f, the system can be reduced using linear model order reduction technique. The only complication is that, after the reduction, the argument of x has to be recovered by the projection $x = Vx_q$ [31,32].

As a typical example Fig. 10 displays the transient simulation and frequency response of the original and reduced models of the microgyroscope being received via usage of the Arnoldi process [33].

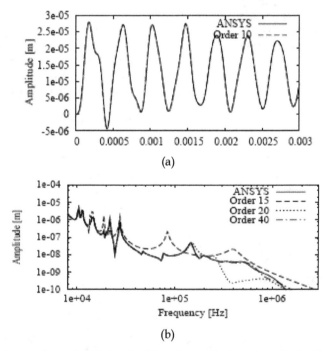

(a)

(b)

Fig. 10. Comparison of transient behavior (a) and transfer functions (b) for the full and reduced microgyroscope models [33].

We see that the solution obtained by reduced model of order 10 is already very close to the true ANSYS excitation (fig.10,a) while the reduced models of order 15 up to 20 show considerable deviations at the high frequency range (fig.10,b). The model with order of 40 shows a perfect match for the lower eigenfrequencies and it is quite closer for higher frequencies, though this is not so important for the gyroscope.

Using Krylov/Arnoldi approach, only a postprocessor to ANSYS is necessary to generate a macromodel in one of the well established model description languages: pure C code, HDL-A, MAST, Modelica and the new standardized VHDL-AMS which are supported by powerful system simulators. Such approach was implemented in *mor4ansys* (pronounced "more for ANSYS") that was developed by IMTEK [33] and which provides the reduced model of order 20 - 30 with an accuracy of a few percents when the dimension of an original FEM is up to 100 000.

3.3 Equivalent-circuit ROM

Taking in to account relations between displacements x, velocities v and accelerations a: $a = dv / dt$, $x = \int v dt$, it is possible to present the equation (2) in the form [34,35]:

$$\frac{d}{dt}(Mv) + Dv + \int Kv dt = F(t) \quad \text{or} \quad \tilde{C}\dot{v} + \tilde{G}v + \tilde{L}v = F(t) , \tag{26}$$

where $\tilde{C} = M, \tilde{G} = D, \tilde{L} = K$ are equivalent matrices of capacitances, conductance and inductances.

$$\tilde{C}, \tilde{G}, \tilde{L}$$

The elements of matrices $\tilde{C}, \tilde{G}, \tilde{L}$ are formed from the elements of the mass, damping and stiffness matrices in the following ways:

$$C_{ij} = -m_{ij}, \quad i, j = 1(1)N, \quad i \neq j; \quad C_{ii} = \sum_{j=1}^{N} m_{ij}, \quad i = 1(1)N; \quad L_{ij} = -1 / k_{ij}, \quad i, j = 1(1)N, \quad i \neq j;$$

$$L_{ii} = 1 / \sum_{j=1}^{N} k_{ij}, \quad i = 1(1)N; \quad G_{ij} = -d_{ij}, \quad i, j = 1(1)N, \quad i \neq j; \quad G_{ii} = \sum_{j=1}^{N} d_{ij}, i = 1(1)N. \tag{27}$$

where N is a number of equations or nodes of the MEMS structure.

In this approach a capacitive-inductive-resistive model of the circuit is built which correctly reflects mass, damping and stiffness matrices. Nodal potentials in this model correspond to the displacement velocities v.

Connected inductors, capacitors and conductors are placed in parallel between each two nodes and each node and ground (for the case i = j), fig.11. Values of these elements are defined by equations (27). Similar approach was used for solving thermo-structural analysis [35].

Displacements x are defined by the current of inductances L_{ii}, which are connected between a node and ground. Such approach suggests significant advantage if the coefficients of the mass and stiffness matrices become time-dependant.

The task of reduction of MEMS model order turns now into reduction of the equivalent RLC circuit size. There are two ways of performing such circuit size decreasing with the minimal accuracy loss:

- sequential excluding of the internal circuit nodes by application of Y-Δ (star-triangle) transformation [36, 39];
- building a macromodel as a four- terminal [40].

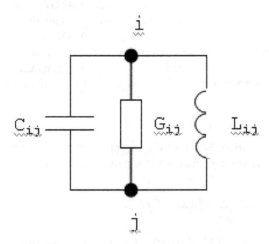

Fig. 11. Circuit element for composing the equivalent-circuit model

3.3.1 Y-Δ transformation based circuit size reduction methods

The essence of the methods based on the Y-Δ transformation consists in the following. Let's i-th node and k its neighbors are located as shown on fig.11. Then the component equation of i-th row will look like

$$Y_i V_i - y_1 V_1 - y_2 V_2 - ... - y_n V_n = 0 \ , \tag{28}$$

where $Y_i = \sum_{j=1}^{k} y_j$.

Let's define V_i as

$$V_i = \left(\sum_{j=1}^{k} y_j V_j \right) \Big/ Y_i \tag{29}$$

and replace V_i in other k equations , that is equivalent to excluding i- th node from a circuit. Then, for example, the equation of the first node transfers to:

$$(\bar{Y}_1 + y_1 - y_1^2 / Y_i) V_1 - \left(\sum_{j=2}^{k} y_1 y_j V_j \right) / Y_i - \sum_{\substack{r=1 \\ r \ne i}}^{k1} y_r V_r = 0 \tag{30}$$

where $\bar{Y}_1 = \sum_{\substack{r=1 \\ r \ne i}}^{k1} y_r$ is a sum of all node conductance excluding i- th node, $k1$ is a number of

nodes being connected to node 1.

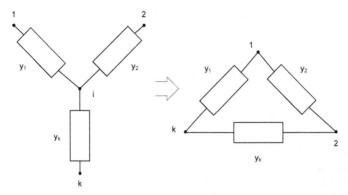

Fig. 12. A working node of the RLC-circuit: conductance are added between node 1 and all neighbors nodes

The equation (30) can be simplified:

$$\left(\bar{Y}_1 + \left(\sum_{j=2}^{k} y_1 y_j\right) / Y_i\right) V_1 - \left(\sum_{j=2}^{k} y_1 y_j V_j\right) / Y_i - \sum_{\substack{r=1 \\ r \neq i}}^{k1} y_r V_r = 0 \qquad (31)$$

Note that it is equivalent to adding k-1 new elements between first node and k-1 former neighbors of i-th node on Fig. 12.

For any two a-th and b-th nodes, which are neighbors to i- th node, elimination of i- th node will add a new element between these two nodes, which is equal to:

$$y_{ab} = (y_a y_b) / Y_i \qquad (32)$$

or in the p-polynomial form, taking into account existence of R, L and C elements, as shown on fig. 11:

$$y_{ab} = \left(g_a + \frac{b_a}{p} + pc_a\right)\left(g_b + \frac{b_b}{p} + pc_b\right) \Big/ \left(G_i + \frac{B_i}{p} + pC_i\right) \qquad (33)$$

where $C_i = \sum_{j=1}^{k} C_j$, $B_i = \sum_{j=1}^{k} B_j$, $G_i = \sum_{j=1}^{k} G_j$ are sums of all the capacitances, reciprocal inductances and conductance connected to i-th node.

In order to simplify (33) two constant time values $\tau_{RCi} = C_i / G_i$ and $\tau_{LCi} = \sqrt{C_i / B_i}$ are introduced for any node in the circuit. The time constant of i-th node is defined as $\tau_i = \max(\tau_{RC}, \tau_{LC})$ and it is considered to be fast if

1. $\tau_i < \tau_{\min} = 2\pi / \omega_{\max}$,

where $\tau_{\min}$ - a time constant which depends on maximal circuit frequency $\omega_{\max}$ being defined by user. So, a fast node will satisfy the following conditions:

$$\omega_{\max} C_i < G_i, \ G_i < B_i / \omega_{\max} \ \text{i} \ \omega_{\max} C_i < B_i / \omega_{\max}.$$

In order to eliminate a fast node from RLC circuit, let us consider the following two cases.

2.   If $\tau_{RCi} \gg \tau_{LCi}$, the equation (33) can be transformed into:

$$y_{ab} = (g_a + pc_a)(g_b + pc_b) / (G_I + pC_i) \tag{34}$$

and its decomposition into Taylor series will look like:

$$y_{ab} = \frac{g_a g_b}{G_i} + p \frac{c_a g_b + c_b g_a}{G_i} + p^2 \frac{4 c_a c_b}{G_i} . \tag{35}$$

For typical R,L,C values (R=0,1÷1000 ohm, L=0,01÷10 nH, C=0,001÷10 pF, ω=0,1÷10 GHz) the contribution of the last term in (35) will be significantly smaller that previous ones. The constant term in (35) gives the value of the conductance, which appears between a- th and b- th nodes during the elimination of i- th node, and the second term in (35) defines the value of capacitance.

3.   If $\tau_{LCi} \gg \tau_{RCi}$, the equation (33) is transformed into:

$$y_{ab} = \frac{1}{p} \frac{b_a b_b}{B_i} + p \frac{c_a b_b + c_a b_b}{B_i} . \tag{36}$$

The first term with $1/p$ in (36) defines the value of the reactive conductance (the inductances' reciprocal value), which appears between a- th and b- th nodes during the elimination of i- th node, and the second term in (36) gives the value of the capacitance.

Final formulas for additional elements between a-th and b-th nodes during the elimination of i- th node for all possible cases are given in tables 1 and 2. All these formulas can be deducted from (35) and (36). The only exception is the case when a- th and b- th nodes are connected to i- th node through a capacitor when $C_{ab} = C_a C_b / C_i$.

In order to eliminate a node from equivalent MEMS circuit with the smallest inaccuracy, the following two criteria should be taken into account:

-    a node should be fast;
-    time constants τ_{LCi}, τ_{RCi} should not be compared (have almost the same values).

In practice, usually lower eigenfrequencies for mechanical systems are most interesting. Therefore, a compromise between accuracy and size of the received circuit models can be reached by proper selection of τ $_{min}$ value. So the algorithm of equivalent-circuit ROM can be described as following:

1.   Time constants should be calculated for all the nodes of the circuit.
2.   All the nodes should be put into a priority queue, sorted by time constants, apart from input, output and ground nodes.
3.   Take the first i- th node in the queue (the one with the smallest time constant $\tau_i < \tau_{min}$).
4.   Find the neighbors k of i- th node.
5.   Eliminated i- th node according to the rules described in tables 1, 2 (depending on τ_{LCi}, τ_{RCi} values).
6.   Time constants in the set k (which consists of the neighbor of i-th node) should be recalculated
7.   Take i- th node from the head of the priority queue and jump to step 3.

Original branch	Substitution	Formula
		$b_{ab}=b_ab_b/B_i$ $C_{ab}=(b_aC_b++b_bC_a)/B_i$
		$b_{ab} = b_ab_b/B_i$ $C_{ab} = b_aC_b/B_i$
		$b_{ab}=b_ab_b/B_i$
		$C_{ab} = b_aC_b/B_i$
		$C_{ab}=b_a{}^*C_b/B_i$
		$C_{ab}=C_aC_b/C$
		$C_{ab}=C_ag_b/G_i$
		$g_{ab}=g_ag_b/G_i$
		$g_{ab}=b_ag_b/G_i$
		$g_{ab}=b_ag_b/G_i$

Table 1. Transformations for $\tau_{LCi} \gg \tau_{RCi}$ case

Original branch	Substitution	Formula
(branch with r_a, i, r_b, C_a, C_b between a and b)	(branch with resistor and C_{ab} between a and b)	$g_{ab}=g_a g_b/G_i$ $C_{ab}=(g_a C_b + {}_b C_a)/G_i$
(branch with r_a, r_b, C_b between a and b)	(branch with resistor and C_{ab} between a and b)	$g_{ab}=g_a g_b/G_i$ $C_{ab}=g_a C_b/G_i$
(branch with C_a, r_b, C_b between a and b)	(branch with C_{ab} between a and b)	$C_{ab}=g_b C_a/G_i$
(branch with C_a, i, r_b between a and b)	(branch with C_{ab} between a and b)	$C_{ab}=g_b C_a/G_i$
(branch with C_a, i, C_b between a and b)	(branch with C_{ab} between a and b)	$C_{ab}=C_a C_b/C$
(branch with r_a, i, r_b between a and b)	(branch with r_{ab} between a and b)	$g_{ab}=g_a g_b/G_i$
(branch with l_a, r_b, C_b between a and b)	(branch with r_{ab} between a and b)	$g_{ab}=b_a g_b/B_i$
(branch with l_a, i, r_b between a and b)	(branch with r_{ab} between a and b)	$g_{ab}=b_a g_b/B_i$
(branch with l_a, C_b between a and b)	(branch with C_{ab} between a and b)	$g_{ab}=b_a C_b/B_i$
(branch with l_a, i, l_b between a and b)	(branch with l_{ab} between a and b)	$g_{ab}=b_a b_b/B_i$

Table 2. Transformations for $\tau_{RCi} \gg \tau_{LCi}$ case

3.3.2 Building MEMS macromodel as a four- terminal

The main idea of this method is to develop a MEMS macromodel as a four-terminal circuit (n-terminal in a general case) as:

$$\begin{bmatrix} I_a \\ I_b \end{bmatrix} = \begin{bmatrix} Y_{aa} & Y_{ab} \\ Y_{ba} & Y_{bb} \end{bmatrix} \begin{bmatrix} U_a \\ U_b \end{bmatrix}, \tag{37}$$

where I_a, U_a are current and voltage at the macromodel input, I_b, U_b are current and voltage at the output. It is suggested to obtain equation (37) directly from the general matrix of the circuit (e.g. admittance matrix Y) according to the expression [40]:

$$Y_{tp} = \frac{1}{\Delta_{aa,bb}} \begin{bmatrix} \Delta_{bb} & -\Delta_{ba} \\ \Delta_{ab} & \Delta_{aa} \end{bmatrix} \tag{38}$$

where Δ_{ij} – an algebraic complement to a_{ij} element of the initial circuit admittance matrix Y, which is equal to a determinant of the matrix being obtained from the matrix Y after eliminating i – th row and j- th column on the crossing of which the given element a_{ij} is situated. In addition the sign of Δ_{ij} is defined by the factor $(-1)^{i+j}$; $\Delta_{aa,bb}$ – redoubled algebraic complement which equals to the determinant of the matrix obtained by eliminating a-th,b-th rows and a-th ,b-th columns and its sign is defined by the factor $(-1)^{a+a+b+b}$.

The necessary algebraic complements can be calculated by the initial matrix inversion procedure, if matrix elements are defined for the selected frequency ω_0, which is fixed by an user:

$$Y^{-1} = \frac{1}{\Delta} \begin{bmatrix} \Delta_{11} & \Delta_{21} \cdots & \Delta_{n1} \\ \Delta_{12} & \Delta_{22} \cdots & \Delta_{n2} \\ \\ \Delta_{1n} & \Delta_{2n} \cdots & \Delta_{nn} \end{bmatrix},$$

where Δ - is the determinant of the initial matrix Y and $\Delta_{aa,bb} = \frac{\Delta_{bb}\Delta_{aa} - \Delta_{ba}\Delta_{ab}}{\Delta}$.

So if numerical values of the inverted Y^{-1} matrix elements are computed:

$$Y^{-1} = \begin{bmatrix} g_{11} & g_{21} \cdots & g_{n1} \\ g_{12} & g_{22} \cdots & g_{n2} \\ \\ g_{1n} & g_{2n} \cdots & g_{nn} \end{bmatrix} \tag{39}$$

it is possible to find parameters of the equivalent four- terminal (38). By choosing elements $g_{aa}, g_{bb}, g_{ab}, g_{ba}$ of the inverted matrix (39) and computing $g_{aa,bb} = g_{aa}g_{bb} - g_{ab}g_{ba}$ it is possible to find parameters of the reduced model (37) from the relations:

$$\begin{aligned} Y_{aa} &= g_{bb}/g_{aa,bb}; \\ Y_{ab} &= -Y_{ba} = g_{ba}/g_{aa,bb}; \\ Y_{bb} &= g_{aa}/g_{aa,bb}. \end{aligned} \tag{40}$$

Since macromodel parameters are determined by sum of the real and imaginary parts:

$$Y_{ij} = a_0 + i\, a_1,\tag{41}$$

it is convenient to represent the circuit macromodel also as a sum of real and imaginary terms:

$$Y_{tp} = \begin{bmatrix} Y_{aa0} & Y_{ab0} \\ Y_{ba0} & Y_{bb0} \end{bmatrix} + i \begin{bmatrix} Y_{aai} & Y_{abi} \\ Y_{bai} & Y_{bbi} \end{bmatrix}.\tag{42}$$

Investigation the expression (42) for different frequencies confirms that for linear circuits at least for frequency range $\omega \le \omega_0$, where ω_0 – frequency at which elements of the inverted matrix (40) are initially defined, real terms of the computed parameters (42) preserve their values and imaginary terms change proportionally to the selected frequency. This confirms the possibility to use macromodel parameters being computed at once frequency in a broad frequency range.

As it was noted earlier, the macromodel form (42) is inconvenient for direct implementation in the circuit simulation software. A real part of Y_{ij} parameters for the stable passive macromodels ($a_0 > 0$) has to be positive while an imaginary part could be negative. If in (41) $a_0 > 0$, $a_1 > 0$ the Y_{ij} component of the four- terminal may be presented by parallel connection of conductance, capacitance and inductance [41]:

$$G_{ij} = \mathrm{Re}\,(Y_{i,j}),\; C_{ij} = k\,\mathrm{Im}\,(Y_{ij})/\,\omega_0,\; L_{ij} = 1/[(k-1)\mathrm{Im}\,(Y_{ij})\,\omega_0].\tag{43}$$

If $a_0 > 0$, $a_1 < 0$ the respective component Y_{ij} may be presented in the same way as parallel connection of conductance, capacitance and inductance but:

$$G_{ij} = \mathrm{Re}\,(Y_{i,j}),\; C_{ij} = (k-1)\,\mathrm{Im}\,(Y_{ij})/\,\omega_0 \;\text{ and }\; L_{ij} = 1/(k\,\mathrm{Im}\,(Y_{ij})\,\omega_0),\tag{44}$$

where k is defined by reactive components values ratio.

So it is possible to select priori the equivalent-circuit macromodel, for example, shown on Fig.13, and to define parameters of its components using expressions (43) and (44).

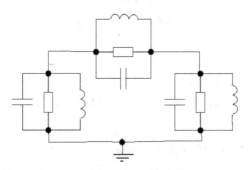

Fig. 13. Equivalent-circuit macromodel of a four-terminal

The important feature of the equivalent circuits MEMS macromodel is a possibility to use optimization procedures for accurate adjustment of the macromodel components values to meet the device characteristics being obtained by FEM model. If a whole MEMS equivalent circuit is rather large it is possible to divide it into some subcircuits and apply transformation (37) to each subcircuit.

Then to combine individual four-terminal circuits into one equivalent MEMS four-terminal circuit taking in account ways of connectivity of different subcircuits (parallel, sequence or mixed) and proper recalculating four-terminal circuits parameters systems (y, z, h, a) [40].

3.3.3 NetALLTED equivalent-circuit ROM subsystem

Equivalent-circuit ROM for MEMS was implemented in the circuit simulation package NetALLTED (**ALL TE**chnologies **Desinger**) which was developed not only for simulation and analysis, but for processing project procedures such as parametric optimization tasks; optimal tolerance assignments; centering availability regions; yield maximization [42]. NetALLTED is widely used for design of Nonlinear Dynamic Systems composed of either/and electronic, hydraulic, pneumatic, mechanical, electromagnetic, and other elements and it is available through the Internet (http://allted.kpi.ua/). ROM developing approach in hand provides more than 99% reduction of elements and node numbers of equivalent- circuit ROM. For example, for the accelerometer only 3 nodes and 6 elements are left from initial 1,883 nodes and 62,826 elements.

Let's consider the example of the equivalent- circuit ROM for the beam working on bending and find its eigenfrequencies [34]. The left end of the beam is fixed motionlessly, right one is free. Force f is applied to the right end perpendicularly to the beam axe.

The initial FEM model was constructed using ANSYS Multiphysics v.10.0 and the beam ANSYS library's BEAM3 finite element with the length of 0.5 um for following beam parameters: L = 25 um; beam cross-section is a square one with height of 3 um and width of 2 um. Beam material properties: coefficient of elasticity E=2 $\cdot 10^{11}$, Pa = 0.2 N/mkm²; Poisson coefficient μ = 0.3; material density ρ = 6 $\cdot 10^3$ kg/m³ = 6 $\cdot 10^{-9}$ mg/mkm³. The initial equivalent beam circuit contains 101 nodes and 314 elements.

The developed equivalent-circuit ROM with 5 nodes and 14 elements is presented on fig.14.

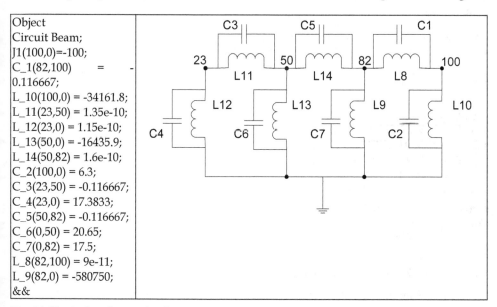

```
Object
Circuit Beam;
J1(100,0)=-100;
C_1(82,100)      =      -
0.116667;
L_10(100,0) = -34161.8;
L_11(23,50) = 1.35e-10;
L_12(23,0) = 1.15e-10;
L_13(50,0) = -16435.9;
L_14(50,82) = 1.6e-10;
C_2(100,0) = 6.3;
C_3(23,50) = -0.116667;
C_4(23,0) = 17.3833;
C_5(50,82) = -0.116667;
C_6(0,50) = 20.65;
C_7(0,82) = 17.5;
L_8(82,100) = 9e-11;
L_9(82,0) = -580750;
&&
```

Fig. 14. Equivalent beam circuit

The two lower eigenfrequencies of the beam are defined by computation. The results of ANSYS Multiphysics frequency analysis as well as the results of the equivalent- circuit ROM simulation for different τ_{min} by NetALLTED [36,43] are given in table 3.

It is obvious that the accuracy of a macromodel obtained with $\tau_{min} = 3*10^{-5}$ is rather high and there are only 5 nodes and 14 elements in the reduced circuit. For more accurate simulation it is possible either to use a reduced circuit obtained with smaller values of τ_{min} (when a size of equivalent circuit ROM increases), or to adapt the received ROM with help of the optimization methods.

| | ANSYS results | ALLTED results | | | |
		Source circuit	Reduced circuit		Optim. circuit
τ_{min}, s	-	-	$5*10^{-6}$	10^{-5}	$3*10^{-5}$	$3*10^{-5}$
Node number	-	101	24	12	5	5
Element number	-	314	76	38	14	14
Reduction by nodes, %	-	-	76,2376	88,1188	95,0495	95,0495
Reduction by elements, %	-	-	75,7962	87,8981	95,5414	95,5414
1st peak, Hz	1336,2	1336,3	1336,1	1334,9	1327	1336,2
2nd peak, Hz	4009,3	4009,3	4009,4	3993,3	3612,1	4009,4
Maximal error, %	-	-	0,01	0,3	9,9	0,003

Table 3. Frequency beam analysis results

The macromodel of the four-terminal for the same beam is being developed in according with (38)-(44) with only 3 nodes and 6 elements that presented on fig.15.

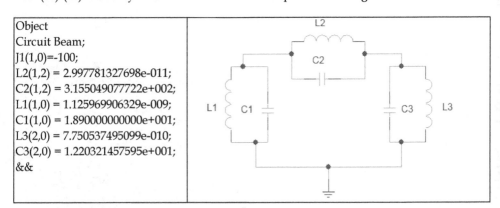

Fig. 15. Equivalent beam circuit being considered as a four- terminal

The results of simulations of the four- terminal equivalent circuit by NetALLTED [36,43] are given in table 4.

	Source circuit	Reduced circuit (as n-ports)
Node number	101	3
Element number	314	6
1st peak, Hz	1336.3	1337.0(0.05%)
2nd peak, Hz	4009.3	4008.9(0.01%)

Table 4. Results of frequency analysis of the four- terminal equivalent circuit for a beam

The advantage of the equivalent circuit approach is obtaining small size of the equivalent reduced circuit as well as the possibility to get required frequencies with a high accuracy, using NetALLTED optimization possibilities (fig.16). The disadvantage is a necessity to make the additional analysis of reduced circuit in order to find the most sensitive elements and take them as variable parameters during optimization procedure.

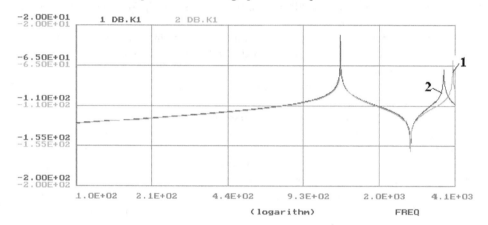

Fig. 16. ROM transfer function before (1) and after (2) optimization.

For example, for reduced circuit, shown at fig.14, four variable elements have been chosen (*L8, L11, L12, L14*) and the objective function *OF ERROR1= F8(1336.3,4009.3/T1,T2)* was constructed which contains the requirement to obtain resonance peaks at frequencies *T1=1336.3* Hz and *T2=4009.3* Hz in according to the ANSYS analysis results (table 3). The objective function contains also current values of frequency response extremes *T1=MAXA(db.K1,100,1600)* and *T2=MAXA(db.K1,1700,4100)*, which are calculated with help of MAXA directive for defining time or frequency when the output (*db.K1* in our case) reaches its maximum value in the specified time or frequency range (*100-1600* Hz). Among available 12 optimization methods being incorporated in NetALLED the Random Search Method (METHOD=40) with search interval reducing has been used to optimize the beam macromodel parameters.

4. MEMS coupled system-level model

The modeling of MEMS provides a very challenging task in modern engineering. This field of research is inherently multiphysics of nature, since different physical phenomena are tightly intertwined at microscale. Typically, up to four different physical domains are usually considered in the analysis of microsystems: *mechanical, electrical, thermal* and *fluidic*. For each of these separate domains, well-established reduced order modeling and analysis techniques are available. However, one of the main challenges in the field of microsystems engineering is to connect models for the behavior of the device in each of these domains to equivalent lumped or reduced-order models without making unacceptably inaccurate assumptions and simplifications and to couple these domains correctly and efficiently.

Micromechanical membrane devices (capacitive pressure transducers, ultrasonic transducers), surface micro machined devices (RF switches, micro optical devices) as well as bulk micro machined devices (accelerometers, inclinometers, laser scanning mirrors) are driven or sensed by nearly parallel electrode pairs in many cases. The motion of these electrodes is strictly normal to the surfaces.

It means that MEMS electrical parts of these MEMS had to be combined with mechanical ones (fig.16). Typically, block-diagram descriptions and lumped-element circuit models for components are connected into a full system. Mostly, this description is used for functional analysis of a design concept.

The coefficients and electrostatic nodal force are obtained from the capacitance-displacement function $C(w)$ of the associated electrode portions and gap space. The function can be input by one of three means [44]:

- as analytical function if the electrode portions make up a plate capacitor geometry with homogeneous intermediate field;
- as polynomial approximation of a function given by data points;
- as data table wherein the element subroutine interpolates values during solution.

The electrostatic forces acting on the movable conductor of the device are included in the model as nonlinear input forces, which are applied on nodes distributed over the conductor surface. Nodes divide the surface into N smaller portions. A lumped force is applied to k- th node at the center of each portion, in its preferential direction of movement x_i. The capacitance C_k between the k-th portion and the fixed electrode of the device is computed, for its undeformed configuration, using an electrostatic analysis

The entity of each force f_k is then approximated as:

$$f_k = \frac{1}{2} \frac{\varepsilon A_k}{(d_0 + w_i^k)^2} (V_p - V_n)^2 \qquad (45)$$

where d_0 is the initial distance between the conductor and the electrode, ε is the relative dielectric permittivity, w_i^k is the displacement of the k- th node along the direction x_i and $(V_p\text{-}V_n)$ is the voltage difference between the structure and the fixed electrode.

The capacitance C_k can be calculated from

$$C_k = \varepsilon A_k/(d - w_i^k). \qquad (46)$$

If C_k varies considerably with the deformation of the structure, then a series of electrostatic computations for different device deflections in its operation range can be performed. The results can be used for extracting the dependency $C_k(w_i^k)$ and calculating the electrostatic force.

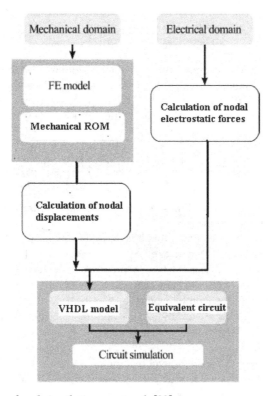

Fig. 16. MEMS system-level simulation approach [33]

According to the assumption that conductors are equipotential, all the nodes connected to a certain conductor are subjected to the same voltage boundary conditions. The total current flowing in the conductor is simply given by the sum of the currents at those nodes:

$$i_k = \frac{d}{dt}[C_k(V_k - V_n)].$$

The nodes used for electrostatic force application can be also used for monitoring the distance between the movable structure and the electrode. When this is equal to the transduction gap, the contact condition is reached and contact forces with the stiffness of the contact K_n, given by equation

$$F_{cont} = K_n(d - gap_{min}), \tag{47}$$

can be applied to nodes.

Computation of the electrostatic forces adds some complexity to the development of the MEMS system-level model, but this is largely compensated by the speed-up simulation of the full model.

There is the special ANSYS' transducer element *TRANS126* which has the possibility to calculate the capacitance of a parallel plate capacitor model (at particular nodes or as a whole) as [47]

$$C(w) = \frac{C_0 d_0}{d_0 - w} = \frac{C_0}{d_0}(1 + \frac{w}{d_0} + \frac{w^2}{d_0^2} + \frac{w^3}{d_0^3} + \frac{w^4}{d_0^4} + ...) \tag{48}$$

where d_o and w are the initial distance and the displacement between the plate.
The element has two nodes; the gap distance is calculated as the sum of the initial displacement and the difference of the nodal displacements in the direction of the element. The force is calculated by equation being similar to (44) so that in the constant voltage case

$$F = \frac{1}{2}\frac{\partial C(w)}{\partial d}V^2 = \frac{C_0 V^2}{2}(\frac{1}{d_0} + 2\frac{w}{d_0^2} + 3\frac{w^2}{d_0^3} + 4\frac{w^3}{d_0^4} + ...) . \tag{49}$$

The element has also contact capabilities (47). It is possible to specify a minimal gap and a spring stiffness K_n for the repelling force.
The drawback of this kind of element is that it is limited to the case where the electrodes are (almost) parallel plates, so that the stroke/capacitance function can be evaluated from a single degree of freedom. But the extensions for rotation plates and for 2D cases were developed. The last one with a triangular shape (element TRANS109) is useful for simulating structures such as comb drivers and optical MEMS, in which capacitance between the device parts is generally a function of a two-directional displacement. TRANS126 and TRANS109 elements enable a huge reduction of the complexity of the system-level simulation.
Let's consider for example the system-level macromodel of an ultrasonic transducer which has two plates with the bottom electrode area A_c and the plate dimension L and which can be presented by nonlinear capacitance:

$$C_{eq} = C_o + (C_{L/2} - C_0)(1 - e^{-\tau t}), \tag{50}$$

where C_0 is the smallest capacitance in the absent of voltage V: $C_0 = \frac{\varepsilon_0 A_c}{d_e}$

$C_{L/2}$ is the largest capacitance when a plate center displacement $w(\frac{L}{2},t)$ is calculated from the ROM macromodel equations (26):

$$C_{L/2} = \frac{\varepsilon_0 A_c}{d_e - w\left(\frac{L}{2},t\right)},$$

where d_e is an equivalent gap ($d_e = d_0 + \frac{d_1}{\varepsilon_1} + \frac{d_{ins}}{\varepsilon_{ins}}$); ε_0 is the absolute dielectric permittivity of the vacuum, ε_1 is the relative dielectric permittivity of the poly silicon, ε_{ins} is the relative dielectric permittivity of the insulator; w is a plate deflection .
The largest value of this capacitance corresponds to the value $w_{max} = d_0$. The electrostatic force acting on the capacitor surfaces is the Coulomb force:

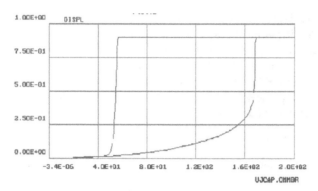

Fig. 17. Capacitive –voltage characteristic an ultrasonic transducer

$$\tilde{F}_{elec} = -\frac{\partial E}{\partial w} = \frac{V_{in}^2}{2}\frac{\partial C_{eq}}{\partial d} = \frac{\varepsilon_0 A_C V_{in}^2}{2\left(d_e - w(\frac{L}{2},t)\right)^2} .$$

As the displacement $w(\frac{L}{2},t)$ reaches the value $w_{max} = d_0$, the hard stop will restrict its further increase, but input voltage V_n and $\tilde{F}_{elec}$ can be further increased. Assume that the input voltage V_n is a superposition of a constant voltage V_{DC} and a time dependent signal $V(t)$ and that $V_{DC} \gg V(t)$. The elastic-plastic properties of the points of contact are simulated by a spring with rigidity K_n and a damper of damping factor b_0. When the beam center moves past w_{max}, it starts interacting with the spring that represents the contact. The damper is introduced to take into account the energy dissipation at the contact. The following equations are used to represent this model:

$$R = \begin{cases} 0 & \text{for } w \le w_{max} \\ K_n(w_{max} - w) - b_0 w' & \text{for } w > w_{max} \end{cases},$$

where R is the interaction force.

The process of interaction simulated by the force R sometimes can be highly sensitive to the values of the model parameters (rigidity and damping factor), especially if $w_{max} \approx d_e$. The interaction can induce high-frequency motions and slow down the rate of convergence considerably. Therefore, a good deal of attention must be paid to accurately modeling and representing this process using experimental data.

If the input voltage is increased more, there will be no equilibrium and the plate collapse takes place. In this case, a hard stop or some other arrangement must be introduced to limit the plate motion. During the plate collapse, the difference between the electrostatic force and the elastic force of the spring will continue to increase. Therefore, when the plate drops down into the hard stop, it is not enough to reduce the input voltage below to release the plate. The input voltage should be reduced more to make the electrostatic force at least equal to the elastic force. Hence, the plate capacitance exhibits the *hysteretic* behavior with respect to voltage change.

But if the insulation layer is rather thick its restrictive effect should be taken into account. If the maximal center displacement W_{max} is equal to initial thickness air gap the moving plate touches the insulator top surface of the electrode when the input V_n voltage reaches the value of V_{max} (V_{max} may be calculated from simulation). But as soon as the input voltage will be decreased under this value the plate will leave the hard stop. So, its capacitance does not demonstrate the hysteretic behavior (fig.17).

Parameter τ in (50) defines an ultrasonic transducer frequency band and can be calculated through the plate displacement $w(\frac{L}{2},t)$ and its velocity $v(\frac{L}{2},t) = w'(\frac{L}{2},t)$, which are defied from ROM equations in the following way:

$$\tau = \frac{1}{3} w(\frac{L}{2},t) / v(\frac{L}{2},t) \tag{51}$$

The coefficient 3 appears in (51) due to the fact that a capacitance recharge to 98% for a time value which is equal approximately 3τ.

It is possible to see two included procedures in according to fig.18: one for development of ROM for an ultrasonic transducer plate, where a deflection and speed of central point`s deflection of transducer plate is calculated for the value V_n, and second - for determine of MEMS system- equivalent capacity value (SLM), using values of plate central point co-ordinates .Then the cycle of calculations recurs whereupon.

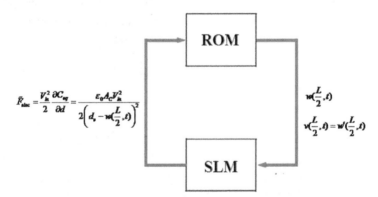

Fig. 18. ROM- system-level model coupled simulations

Instead of using two sequence procedures mentioned above it is possible using functional possibilities of the circuit simulator NetALLTED to built a single system-level equivalent circuit model for an ultrasonic transducer by introducing directly into the equivalent- circuit ROM of mechanical MEMS part the additional arbitrarily connected element (a Depended Source) with an informative function which is determined by equation (50) [42]. Optimization procedures of NetALLTED allow getting the desirable values of this transducer capacity and through it to get a desirable value of output signal of an ultrasonic transducer system-level model by the changing ROM parameters, which, in turn, are depended upon an ultrasonic transducer construction sizes and used material properties.

Providing a single representation of a MEMS operating in multiple physical domains, the electrical circuit approach is very convenient. Moreover, powerful mathematical techniques and circuit simulation programs are available for solving design tasks. It is possible to develop a library of schematic model for different MEMS elements and then use their combinations to build a system-level macromodels for entire rather complicated MEMS constructions.

5. Conclusion

In this chapter, the methods and issues encountered in the development of MEMS macromodels at the system level have been presented. System level modeling is the highest and most abstract level of modeling. This level requires various devices` linking of MEMS *component level models* – both electronic and micromechanical – into a micro-electro-mechanical system. *System-level models* of MEMS components are needed to allow a fast and sufficiently exact investigation of their behavior to simulate entire MEMS.

Starting point for the extraction of a reduced order model (ROM or a macromodel) is already its description with a large ODE system, which is typically derived using physical modeling techniques based on Finite Element Method (FEM) which is rather time consuming. Macromodels application allows the extraction of lower order ODE system that reproduces the input/output behavior with good accuracy. Particular attention has been posed in the chapter on the possibility to get a macromodel circuit presentation.

There are special methods for generating ROM for MEMS components and entire MEMS based on FEM descriptions. To derive macromodels of smaller sizes different approaches (*Modal decomposition, Moment matching, Equivalent circuit presentation*) were developed. Usage of the reduced MEMS components models allows applying successfully modern circuit simulators in workflow for MEMS design on system level.

Three automatic procedures to generate device reduced order macromodels, being based on full FEM/FDM models, were demonstrated in this chapter. Two of them are suitable for simulators with possibilities to get input information in the equation forms (ODE or OAE). The third one in opposite produces macromodels in circuit presentation and so it is more suitable for circuit simulators. The Modal ROM approach is based on using natural (modal or resonant) frequencies of MEMS structure and it is spread mostly in the USA and Asia. The Moment matching ROM approach is based on using the Krylov subspace for transfer functions and it is popular in Western Europe and Asia. The Equivalent circuit ROM approach is based on using a capacitive-inductive-resistive circuit model for mass, damping and stiffness matrices and it is used mostly in Eastern Europe. It is worth to notice that the Modal ROM approach requires some full ANSYS runs to perform a proper orthogonal decomposition during basic functions determination in opposite to the Moment matching and the Equivalent circuit ROM approaches for which it is enough to use ANSYS only for FEM model matrices building.

It seems to be interesting and perspective trying to combine mentioned approaches, for example, to start with Krylov/Arnoldi reduction of ODE dimension, then to build the proper equivalent circuit for obtained ODE systems and finally to apply Y/Δ transformation or n-port transformation for further reducing macromodel order.

6. References

[1] S. D. Senturia, "CAD challenges for microsensors, microactuators, and microsystems," *Proc. IEEE*, vol. 86, pp. 1611–1626, 1998.

[2] E. B. Rudnyi and J. G. Korvink, "Model order reduction for large scale engineering models developed in Ansys." *Lect. Notes Comput. Sc.*, vol. 3732, pp. 349–356, 2006.

[3] M. G.Mand G. Ostergaard, "Electro-mechanical trasducer for MEMS analysis in Ansys," in *Proc. Int. Conf. on Modeling and Simulation of Mycrosystems (MSM)'99*, pp. 270–273, 1999.

[4] W. Z. Lin, K. H. Lee, S. P. Lim, and Y. C. Liang, "Proper orthogonal decomposition and component mode synthesis in macromodel generation for the dynamic simulation of a complex MEMS device," *Journal of Micromechanics and Microengineering*, vol. 13, no. 5, pp. 646–654, 2003.

[5] A.I.Petrenko," Design Methology and Workflow for MEMS Design ", *Proc. MEMSTECH'2011, 11-14 May, Polyana-Svalyava (Zakarpattya), UKRAINE*, pp.12-15, 2011.

[6] Peter Avitabile. "Experimental Modal Analysis- A Simple Non-Mathematical Presentation", *Modal Analysis and Control Lab., University of Massachusetts Lowell*, 16 p.,2001.

[7] Gunner C. Larsen, Morten H. Hansen, Andreas Baumgart, Ingemar Carl. " Modal Analysis of Wind Turbine Blades", *Riso–R–1181(EN), ISBN 87–550–2697–4 (Internet), Riso National Laboratory, Roskilde*, Denmark, 72 p., February 2002.

[8] W. Z. Lin, S. P. Lim, and Y. C. Liang, "Proper orthogonal decomposition and component mode synthesis in macromodel generation for the dynamic simulation of a complex MEMS device," *J. Micromech. Microeng.*, vol. 13, pp. 646–654, 2003.

[9] H. M. Park and D. H. Cho, "The use of the Karhunen-Lo`eve decomposition for the modeling of distributed parameter systems," *Chem. Eng. Sci.*, vol. 51, no. 1, pp. 81–98, 1996.

[10] Y. C. Liang, W. Z. Lin, H. P. Lee, S. P. Lim, K. H. Lee, and D. P. Feng, "A neural-network-based method of model reduction for the dynamic simulation of MEMS," *Journal of Micromechanics and Microengineering*, vol. 11, no. 3, pp. 226–233, 2001.

[11] Elmer S. Hung and Stephen D. Senturia." Generating Efficient Dynamical Models for Micro-electro-mechanical Systems from a Few Finite-Element Simulation Runs", *Journal of Microelectromechanical Systems*, vol. 8, no. 3, September 1999.- pp.280-289.

[12] Lynn D. Gabbay, Jan E. Mehner, and Stephen D. Senturia. "Computer-Aided Generation of Nonlinear Reduced-Order Dynamic Macromodels—I: Non-Stress-Stiffened Case", *Journal of Microelectromechanical Systems,,* vol. 9, no. 2, June 2000.- pp.262-269.

[13] Jan E. Mehner, Lynn D. Gabbay, and Stephen D. Senturia. "Computer-Aided Generation of Nonlinear Reduced-Order Dynamic Macromodels—II: Stress-Stiffened Case", *Journal of Microelectromechanical Systems*, vol. 9, no. 2, June 2000.- pp.270-278.

[14] Schlegel, M.; Bennini, F.; Mehner, J.; Herrmann, G.; Muller, D.; Dozel, W.: "Analyzing and Simulation of MEMS in VHDL-AMS Based on Reduced Order FE-Models", *IEEE Sensors 2003*, Second IEEE International Conference on Sensors, Toronto, Canada, 2003.

[15] Bennini, F.; Mehner, J.; Dotzel, W. " Computational Methods for Reduced Order Modeling of Coupled Domain Simulations", *Proc. of 11 International Conference on Solid State Sensors and Actuators (Transducers 01)*, Germany, 2001.

[16] M. Schlegel, F. Bennini, J. E. Mehner, G. Herrmann, D. Mueller, and W. Doetzel, "Analyzing and simulation of MEMS in VHDL-AMS based on reduced-order FE models" , *IEEE Sensors J.*, vol. 5, no. 5, 2005.

[17] J. Chen and S. M. Kang, "An algorithm for automatic model reduction of nonlinear MEMS devices," in *Proc. IEEE Int. Symp. Circuits and Syst.*, pp. 445–448, 2000.

[18] E. S. Hung, Y.-J. Yang, and S. D. Senturia, "Low-order models for fast dynamical simulation of MEMS microstructures" , *Digest of Technical Papers, IEEE Int. Conf. on Solid-State Sensors Actuators and Microsystems (Transducers'97)*, pp. 1101–1104, 1997.

[19] L. H. Feng, "Review of model order reduction methods for numerical simulation of nonlinear circuits," *Appl. Math. Comput.*, vol. 167, no. 1, pp. 576–591, 2005.

[20] M. I. Younis, E. M. Abdel-Rahman, and A. Nayfeh, "A reduced order model for electrically actuated microbeam-based MEMS" , *IEEE J. Microelectromech. S.*, vol. 12, no. 5, pp. 672–680, 2003.

[21] A. H. Nayfeh, M. I. Younis, and E. M. Abdel-Rahman, "Reduced-order models for MEMS applications," *Nonlinear Dynamics*, vol. 41, no. 1–3, pp. 211–236, 2005.

[22] Chun-Hsu Ko ,Jin-Chern Chiou." Fuzzy Macromodel for Dynamic Simulation of Microelectromechanical Systems" , *IEEE Trans.on Systems, Man and Cybernetics – Part A: Systems and Humans*, vol. 36, no. 4, pp.823-830, JULY 2006

[23] Pu Li, Yuming Fang. "A Wavelet Interpolation Galerkin Method for the Simulation of MEMS Devices under the Effect of Squeeze Film Damping", Hindawi Publishing Corporation, *Mathematical Problems in Engineering*, Article ID 586718, pp.1-25,2010.

[24] Z. Bai, "Krylov subspace techniques for reduced-order modeling of large-scale dynamical systems," *Applied Numerical Mathematics*, vol. 43, pp. 9–44, 2002.

[25] Zhaojun Bai, Daniel Skoogh," Krylov Subspace Techniques for Reduced-Order Modeling of Nonlinear Dynamical Systems", 2002

[26] Jan Lienemann. "Complexity reduction techniques for advanced MEMS actuators simulation," *Ph.D. dissertation, Albert-Ludwigs Universit at Freiburg im Breisgau*, 298 p.,2006.

[27] A. Odabasioglu, M. Celik, L. T. Pileggi, PRIMA: Passive reduced-order interconnect macromodeling algorithm, *IEEE Trans Comput Aid D*, 17 (8) pp. 645–654,1998.

[28] Lihong Feng. "Review of model order reduction methods for numerical simulation of nonlinear circuits", *Applied Mathematics and Computation, vol. 167* , pp. 576–591,2005.

[29] Z. J. Bai and Y. Su, "SOAR: A second-order Arnoldi method for the solution of the quadratic eigenvalue problem," *SIAM J. Matrix Anal. A*, vol. 26, no. 3, p. 640659, 2005.

[30] Z. Bai, P. M. Dewilde, and R. W. Freund, "Reduced-order modeling," *Numerical Analysis*, vol. 02, pp. 1–59, 2002.

[31] M. Rewienski and J.White, "A trajectory piecewise-linear approach to model order reduction and fast simulation of nonlinear circuits and micromachined devices," *IEEE Trans. Comput.Aid. D.*, vol. 22, p. 155170, 2003.

[32] J. Chen, Sung-Mo Kang, Jun Zou,Chang Liu and José E. Schutt-Ainé. "Reduced-Order Modeling of Weakly Nonlinear MEMS Devices with Taylor-Series Expansion and Arnoldi Approach", *IEEE J. Microelectromech. S*, vol. 13, no. 3, pp.441-446, JUNE 2004.

[33] L. Del Tin, J. Iannacci, R. Gaddi, A. Gnudi, E. B. Rudnyi, A. Greiner and J. G. Korvink. "Non Linear Compact Modeling of RF-MEMS Switches by means of Model order rerduction", *Solid-State Sensors, Actuators and Microsystems International Conference, 2007. TRANSDUCERS 2007. Lyon*, 10-14 June 2007, pp.635 – 638.

[34] A.Petrenko, V.Ladogubets, O.Beznosyk, O.Finogenov, "Using Optimization Procedures to Calculate Parameters of MEMS Macromodels", *Proc. CADSM'2009, 24-28 February, Polyana-Svalyava (Zakarpattya), UKRAINE*, pp.511-514, 2009.

[35] Hsu J.L., Vu-Quoc . A rational formulation of termal circuit models for electrothermal simulation. -part 1: finite element method. *IEEE Trans. Circuits & Systems – I:Fund.Theory &Appl.*, 43, 9, pp.721-732, 1996.

[36] Beznosyk O., Ladogubets V., Finogenov O., Tchkalov O. "Using circuit design software to simulate microelectromechanical components // MEMSTECH 2008,

IV-th International Conference on Perspective Technologies and Methods in MEMS Design, Polyana,Ukraine, pp.130-133, 2008.

[37] H. Tilmans, "Equivalent circuit representation of electromechanical transducers: I. lumped-parameter systems," *IEEE J. Electromicromech. Syst.*, vol. 6, no. 1, pp. 157–176, Mar. 1996.

[38] T. Veijola, H. Kuisma, and J. Lahdenper.a, "Dynamic modelling and simulation of microelectromechanical devices with a circuit simulation program," in *Proc. Int. Conf. on Modeling and Simulation of Mycrosystems (MSM)'98*, pp. 245–50, 1998.38. Sheehan JB.N.

[39] TICER: Realizable Reduction of Extracted RC Circuits // *Digest of Technical Papers – IEEE/ACM Proc. of ICCAD*, pp. 200-203, 1999.

[40] Petrenko A., Sigorsky V., "Algorithmic analysis of electronic circuits", *Western Periodical Corp., San Francisco*, 618 p, 1975.

[41] Petrenko A.I.," RLC – circuits models size reduction ", *Proc. CADSM'05, Lviv-Polena*, p.3-8, 2005.

[42] Petrenko A.I., Ladogubets V.V., Tchkalov V.V., Pudlowsky Z.J. ALLTED - a Computer - Aided System for of Electronic Circuit Design, *UICEE.(UNESCO), Melbourne*, 205 p, 1997.

[43] Beznosyk O., Finogenov O., Ladogubets V. Presentation of a System of Ordinary Differential Equations as an Equivalent Electrical Circuit.- // *Perspective Technologies and Methods in MEMS Design : VI-th International Conference MEMSTECH'2010, 20-23 April 2010*, Lviv-Polyana, Ukraine: proc. – Lviv : Publishing House Vezha&Co, 2010. – P. 116–120.

[44] J.Wibbeler, J.Mehner, F.Vogel, F. Bennini. " Development of ANSYS/Multiphysics Modules for MEMS by CAD-FEM GmbH", *19-th CAD-FEM Users' Meeting 2001, International Congress on FEM Technology, Berlin, Potsdam, October* , 2001, pp.1-10.

[45] Sven Reitz, Jens Bastian, Joachim Haase, Peter Schneider, Peter Schwarz." System Level Modeling of Microsystems using Order Reduction Methods", *Journal Analog Integrated Circuits and Signal Processing, vol.37, Issue 1*, 2003, pp.7-16.

[46] L. Del Tin, R. Gaddi, E. B. Rudnyi, A. Gnudi, A. Greiner, J. G. Korvink, "Nonlinear compact modeling of RF-MEMS switches by means of model order reduction", *Proc. 14th International Conference on Solid-State Sensors, Actuators and Microsystems (Transducer'07), 10-14 June, Lyon* (France).2007.

[47] M. G.Mand G. Ostergaard, "Electro-mechanical trasducer for MEMS analysis in Ansys," in *Proc. Int. Conf. on Modeling and Simulation of Mycrosystems (MSM)'99*, 1999, pp. 270–273.

[48] J. Iannacci, L. Del Tin, R. Gaddi, A. Gnudi and K. J. Rangra,"Compact Modeling of a MEMS Toggle-Switch based on Modified Nodal Analysis", Symposium on Design, Test, Integration and Packagingof MEMS/MOEMS (DTIP) 2005, Montreux, Switzerland, 0103 June 2005.

[49] Peter Schwarz. "Microsystem CAD: From FEM to System Simulation", *Proc. Intern. Conf. "Simulation of Semiconductor Processes and Devices" (SISPAD98), Leuven, 2-4. Sept. 1998*, pp.141-148, Springer, Wien 1998.

[50] Xuan F. Zha, Ram D. Sriram. " Information and Knowledge modeling for computer supported Micro-electro-mechanical systems design and development", *Proceedings of ASME Design Engineering Technical Conference (DETC;2005) September 24-28, 2005, Long Beach, California, USA*, pp.1-11.

Permissions

The contributors of this book come from diverse backgrounds, making this book a truly international effort. This book will bring forth new frontiers with its revolutionizing research information and detailed analysis of the nascent developments around the world.

We would like to thank Dr. Nazmul Islam, for lending his expertise to make the book truly unique. He has played a crucial role in the development of this book. Without his invaluable contribution this book wouldn't have been possible. He has made vital efforts to compile up to date information on the varied aspects of this subject to make this book a valuable addition to the collection of many professionals and students.

This book was conceptualized with the vision of imparting up-to-date information and advanced data in this field. To ensure the same, a matchless editorial board was set up. Every individual on the board went through rigorous rounds of assessment to prove their worth. After which they invested a large part of their time researching and compiling the most relevant data for our readers. Conferences and sessions were held from time to time between the editorial board and the contributing authors to present the data in the most comprehensible form. The editorial team has worked tirelessly to provide valuable and valid information to help people across the globe.

Every chapter published in this book has been scrutinized by our experts. Their significance has been extensively debated. The topics covered herein carry significant findings which will fuel the growth of the discipline. They may even be implemented as practical applications or may be referred to as a beginning point for another development. Chapters in this book were first published by InTech; hereby published with permission under the Creative Commons Attribution License or equivalent.

The editorial board has been involved in producing this book since its inception. They have spent rigorous hours researching and exploring the diverse topics which have resulted in the successful publishing of this book. They have passed on their knowledge of decades through this book. To expedite this challenging task, the publisher supported the team at every step. A small team of assistant editors was also appointed to further simplify the editing procedure and attain best results for the readers.

Our editorial team has been hand-picked from every corner of the world. Their multi-ethnicity adds dynamic inputs to the discussions which result in innovative outcomes. These outcomes are then further discussed with the researchers and contributors who give their valuable feedback and opinion regarding the same. The feedback is then collaborated with the researches and they are edited in a comprehensive manner to aid the understanding of the subject.

Apart from the editorial board, the designing team has also invested a significant amount of their time in understanding the subject and creating the most relevant covers. They scrutinized every image to scout for the most suitable representation of the subject and create an appropriate cover for the book.

The publishing team has been involved in this book since its early stages. They were actively engaged in every process, be it collecting the data, connecting with the contributors or procuring relevant information. The team has been an ardent support to the editorial, designing and production team. Their endless efforts to recruit the best for this project, has resulted in the accomplishment of this book. They are a veteran in the field of academics and their pool of knowledge is as vast as their experience in printing. Their expertise and guidance has proved useful at every step. Their uncompromising quality standards have made this book an exceptional effort. Their encouragement from time to time has been an inspiration for everyone.

The publisher and the editorial board hope that this book will prove to be a valuable piece of knowledge for researchers, students, practitioners and scholars across the globe.

List of Contributors

Xing Chen, Dafu Cui, Haoyuan Cai, Hui Li, Jianhai Sun and Lulu Zhang
State Key Lab. of Transducer Tech., Inst. of Electronics, Chinese Academy of Sciences, China

Wen Li
Michigan State University, USA

Damien C. Rodger, James D. Weiland and Mark S. Humayun
University of Southern California, USA

Yu-Chong Tai
California Institute of Technology, USA

Wentai Liu
University of California, Santa Cruz, USA

Ioana Voiculescu
City College of New York, USA

Anis N. Nordin
International Islamic University, Malaysia

Nazmul Islam and Saief Sayed
MEMS/NEMS Lab, The University of Texas at Brownsville, USA

Tong Guo
Tianjin University, P.R. China

Long Ma
Civil Aviation University of China, P.R. China

Yan Bian
Tianjin University of Technology and Education, P.R. China

Y. F. Peng and Y. B. Guo
Xiamen University, China

Bin Tang and Kazuo Sato
Nagoya University, Japan

Anatoly Petrenko
Systems Design Department, National Technical University of Ukraine, Ukrain

Printed in the USA
CPSIA information can be obtained
at www.ICGtesting.com
JSHW011358221024
72173JS00003B/338